IMPROVEMENT TRENDS FOR INTERNAL COMBUSTION ENGINES

Edited by **Bilge Albayrak Ceper**
and **Melih Yıldız**

Improvement Trends for Internal Combustion Engines

http://dx.doi.org/10.5772/66235
Edited by Bilge Albayrak Ceper and Melih Yıldız

Contributors

Joanna Czarnocka, Małgorzata Odziemkowska, Katarzyna Wawryniuk, Ornella Chiavola, Giancarlo Chiatti, Erasmo Recco, Wladyslaw Mitianiec, Chao Cheng, Piotr Lijewski, Lennox Zumbe Siwale, Abdul Rashid Abdul Aziz, Firmansyah -, Morgan Heikal, Ezrann Zharif Zainal Abidin, Naveenchandran Panchatcharam

Published by InTech

Janeza Trdine 9, 51000 Rijeka, Croatia

Publishing Process Manager Nina Kalinic
Technical Editor SPi Global
Cover InTech Design team

First published March, 2018
Printed in Croatia
Legal deposit, Croatia: National and University Library in Zagreb

Additional hard copies can be obtained from orders@intechopen.com

Improvement Trends for Internal Combustion Engines, Edited by Bilge Albayrak Ceper and Melih Yıldız
p. cm.
Print ISBN 978-953-51-3891-4
Online ISBN 978-953-51-3892-1

For EU product safety concerns:
IN TECH d.o.o., Prolaz Marije Krucifikse Kozulić 3, 51000 Rijeka, Croatia,
info@intechopen.com or visit our website at intechopen.com.

Meet the editors

Bilge was born in Kayseri in 1979. She completed her high school education in 1996 in Kayseri. She received her BSc degree from Energy Division of Mechanical Engineering Department in Erciyes University in 2001 and her MSc degree in 2003 and PhD degree in 2009 from the Institute of Natural Science from the same university. She started her job as a research assistant at the Mechanical Engineering Department in Erciyes University in 2001. She studied biogas alternative fuels at CNRS (The National Center for Scientific Research) in 2009. She is now an associate professor doctor in Mechanics in the Department of Mechanical Engineering, Erciyes University. She has a lot of national and international articles in SCI index. She worked in many research projects about internal combustion engines, supported by TUBITAK (The Scientific and Technological Research Council of Turkey) and Erciyes as a supervisor. Her research interests involve the following: internal combustion engine, alternative fuels, computational fluid dynamics, and energy.

Research Assistant Melih Yıldız was born in Sivas, Turkey, in 1979, and he graduated in Mechanical Engineering at Erciyes University in 2002. He worked in a private company that produced heating equipment, such as steam boiler and heat exchanger, as a manufacturing director for seven years. He was involved in many projects such as the establishment of heating plants in that period. He started working at Iğdır University, Turkey, as a research assistant in 2011. Since 2012, he has been a research assistant at Erciyes University. He received his MSc degree in Mechanical Engineering in 2013 from Erciyes University. He went on to complete his PhD thesis that was related to low-temperature combustion engines. He has experience in the establishment of engine experimental setups and took part in many research projects about internal combustion engines. His research area consists of internal combustion engines, alternative fuels, computational fluid dynamics, heat transfer, and thermodynamics.

Contents

Preface

This book on internal combustion engines consists of eight chapters from the research activities through the wide range of engine performances, emission parameters, and alternative fuels. The first three sections are related to the use of alternative fuels and the stability changes of model fuel blends used in internal combustion engines. The following two sections deal with advanced combustion technology and combustion control strategy. Emission measurement methodology is given in the sixth section. The last two sections present modelling studies for modern engine systems. I believe that this book contributes to the readers to follow current improvement trends for internal combustion engine and it can be used by postgraduate students.

Assoc. Prof. Dr. Bilge Albayrak Ceper
Department of Mechanical Engineering
Erciyes University
Melikgazi/Kayseri
Turkey

Research Assistant Melih Yıldız
Erciyes University
Melikgazi/Kayseri
Turkey

Chapter 1

Study of Stability Changes of Model Fuel Blends

Małgorzata Odziemkowska, Joanna Czarnocka and
Katarzyna Wawryniuk

Additional information is available at the end of the chapter

http://dx.doi.org/10.5772/67056

Abstract

Fuel during long-term storage, even at low temperatures, undergoes oxidation to various products, which are further converted into high molecular compounds precipitating from the liquid. The stability of the stored fuel depends on its chemical composition and storage conditions. Chemical conversions associated with the degradation of fuel are not well recognized, so it is difficult to predict the storage duration of such fuel and control the rate of ageing. Because of the variety of hydrocarbon composition of commercial fuels resulting from the diversity of crude oil and oil processing methods, surrogate fuels are used for simulating behaviour of commercial fuels in the various tests. In this paper, the type and dose of model components of fuel were chosen. To determine the degradation degree of model blends, few tests of accelerated ageing were selected. These methods were used to determine induction period, content of deposits and resistance to oxidation. Effect of inhibiting and catalyzing compounds on fuel oxidation was verified. An infrared spectroscopy analysis of the oxidized model blends was also performed. The results of this study showed slight changes of selected properties of the surrogate fuels. The most significant changes occurred for the model blends containing substances initiating radical autoxidation reactions and bio-components.

Keywords: model fuels, hydrocarbons, oxidative stability, induction period, accelerated ageing test

1. Introduction

Chemical stability of fuel, in terms of resistance to changes during storage and distribution, is an important parameter determining its quality. Processes deteriorating the fuel properties include, among others, oxidation. The oxidation products are chemically active substances and undergo further transformations to non-volatile macromolecular substance (resin), and the particulates precipitate in the form of deposits/sediments. Fuels, even under stable storage conditions, can undergo autoxidation, where the hydrocarbon molecules react with atmospheric oxygen or with themselves [1].

There is an abundance of literature on fuel instability [2–13]. Nevertheless, reactions related with the degradation of the fuel are not well researched, so it is difficult to predict the rate and direction of changes of the fuel properties during storage and distribution. In addition, the diversity of the composition of hydrocarbon in commercial fuel species resulting from the variety types of processed crude oil and refinery technologies makes it difficult to determine the mechanisms of these changes. There are many theories describing the mechanisms of oxidation of petroleum products [14].

Generally, it is assumed that the fuels are oxidized by a radical mechanism, whereby the hydrocarbons are involved in radical chain reactions in the following sequence: initiation, propagation, branching and chain termination. The initiation reaction known as the initiation of an oxidative chain reaction begins with an attack of oxygen molecules from the air on the C-H bond in the hydrocarbon molecules. As a result, alkyl radicals (the radicals react with oxygen to hydroperoxides, which are aggressive oxidants) are formed. Decomposition products of peroxides are active alkoxy and hydroxyl radicals, which in the following reaction steps detach another hydrogen atoms from the hydrocarbons (propagation phase and chain branching). The last phase of hydrocarbon oxidation is termination, which is the completion of the reaction, where due to the recombination of hydrocarbon and peroxide, free radicals become deactivated creating non-radical products, for example, alcohols, ketones and acids, which can undergo further transformations to macromolecular substances. Among others, Denisov [15] described in detail oxidation of hydrocarbons in the liquid phase. However, we still do not know how an oxidation chain, in the absence of oxidation promoters such as sunlight, a source of free radicals and metal ions, is formed.

It is considered that the chemical structure of hydrocarbons has an influence on the rate of oxidation processes of petrol. The studies presented in the article [5] have shown that in petrol the tendency to produce resins depends on the amount and type of unsaturated hydrocarbons. In the case of petrol fractions from the cracking process, the presence of active radicals and trace amounts of metal ions originating from the catalyst was found. Pereira et al. [6] proved that not all of the olefins presented in the fuel are transformed equally to resins. Among tested olefins, those that formed the secondary allyl radicals (2,4-hexadiene and cyclohexene) had the largest share in the formation of macromolecular substances. These compounds are classified, respectively, to the unsaturated hydrocarbons with conjugated bonds and cyclic olefins. The considerations regarding to the stability of allyl and alkyl radicals and autoxidation reaction mechanism confirming the radical chain reactions were presented. Research of effect of ethanol and copper presence on the stability of petrol [7] showed that the addition of alcohol does not deteriorate fuel stability, whereas copper in the system is undesirable because it accelerates the oxidation processes.

However, in the case of diesel fuels, chemical instability is caused by the presence in the fuel compounds containing nitrogen and sulphur, reactive olefins and organic acids which are precursors of macromolecular structures having a limited solubility. One of the well-recognized mechanisms generating insoluble precipitates in the diesel fuel is conversion of phenalenes and indoles to indole/phenalene salt complexes, described by Marshman and Davia [8]. An acidic environment is conducive to the reactions. Phenalenes are formed as a result of the oxidation of active olefins, and indoles are a natural component of the fuel. The organic acids (necessary to catalyze the reaction) are typically presented in the fuel components or are formed by oxidation of the mercaptans to sulfonic acids. Li et al. [10] conducted research on the thermo-oxidative stability of aviation fuel. Oxidation processes were followed by determination of hydroperoxides and FTIR spectral analysis. The results confirmed the theory presented by Zabarnick [11] that the fuel oxidation occurs according to the mechanism of free radical chain reactions in which the hydroperoxides can be considered as intermediates for subsequent reactions of hydrocarbon autoxidation. During the tests it was observed that the fuel that contains more polar components is more readily oxidized. Normally, the presence of polar components is the main factor of instability.

The complexity of the hydrocarbon fuel oil composition is a difficult in-depth analysis of the degradation process of the fuel components and consequently the finished fuel which is a mixture of hundreds of compounds. The chemical composition of the fuel determines the physicochemical properties and influences on the engine operating characteristics [16]. Commercial fuels, depending on a producer, may differ from each other in terms of composition and physicochemical properties. Fuel oxidation and its stability issues were studied by numerous scientists for simple systems using pure hydrocarbons, representative for each fuel type or model mixtures. A key decision about the composition of the model fuel is a selection of the individual compounds—most often pure components representing a hydrocarbon groups (it is assumed that their chemical interactions are typical of other similar chemical structures) are selected for the composing process.

Gernigon et al. [17] conducted a study aimed at recognition of the oxidation mechanisms of aviation fuel. The research was performed for model mixtures containing such mono-compounds as n-dodecane, 1,3-diisopropylbenzene, cyclohexylbenzene and 2,2,4,4,6,8,8-heptamethylnonane in characteristic proportions of the natural composition of the fuel. Studies of the decomposition kinetics of hydrocarbons showed that after 48 hours, test diminished from 40 to 80% of each of mono-compounds. For all hydrocarbons, two stages of degradation were recorded: the first is the rapid decrease in the hydrocarbon content during the first hour of the experiment; a second step, after approx. 8 hours, is characterized by slow 'consumption' of hydrocarbon and has an almost linear behaviour. IR spectrum analysis indicated that the new structures are formed, because with increasing time, the degradation of characteristic band intensity, for the compounds formed by the oxidation (OH, C-O, C=O), increases. The antioxidants used during the tests slowed down the hydrocarbons degradation.

Li et al. [10] used n-cetane of 99% purity as a reference fuel for spectroscopic studies of oxidized fuels. Natelson et al. [18] conducted a comparative study of aviation fuel and diesel reactivity with respect to mixtures of chemical compounds that have been selected as a model fuel (surrogate of diesel and aviation fuel). The studies described herein have focused on three substitutes, from three hydrocarbon groups (paraffins, naphthenes, aromatics), which are in the aircraft fuel and diesel. In order to maintain the small number of elements for modelling, one component from each of these groups was selected. N-decane ($C_{10}H_{22}$) was the representative of paraffins, n-butyl cyclohexane ($C_{10}H_{20}$) was representative of naphthenes, and n-butylbenzene ($C_{10}H_{14}$) was a representative of aromatics. The surrogates have been used in respective proportions, depending on the structure and key features of the real fuel. The surrogate reactivity was compared to the average of ordinary aviation fuel and commercial diesel used in the United States.

The authors of following research conducted the study on the stability of diesel fuel during long-term storage and the results presented in the work [1]. In the case of petrol, after 6-month storage, in the IR spectra, slight change in the chemical composition was found. The oxidation-promoting substances, i.e. cyclic olefins and hydroperoxide intentionally added to the stored fuel, slightly influenced on intensification of transformation process. The study of induction period and potential resins content showed the potential adverse changes in petrol stability, which were poorly correlated with changes demonstrated by IR spectroscopy. However, in diesel fuel samples after 3–4 months of storage, there was a significant increase in the content of sediments. The IR spectra confirmed the changes in the chemical composition of the fuel. In the stored samples appeared degradation products of hydrocarbons, containing functional groups characteristic for acids, aldehydes, ketones and phenols, conducive to further reactions leading to the formation of indole/phenalenes.

Aforementioned studies were carried out for commercial fuels containing hundreds of chemical components of different elemental compositions and structures. Standard tests determining the petrol stability which include induction period [19] and resins content [20] are laborious and time-consuming. Also, the tests do not allow for early detection of changes in the fuel stability, which are visible on the IR spectra. The situation is similar in the assessment of stability of middle distillates; the scientists have much more methods to evaluate this parameter [21–24]. The authors have attempted to assess the stability of petrol using a quick and simple method of induction period at the micro level. To eliminate the influence of a large number of compounds presented in petrol and the interactions between them, the study was carried out for the model mixture replacing commercial petrol. Furthermore it was checked how the presence of the chemically unstable component (cyclohexene) changes the oxidation stability of model petrol. The induction period for the model mixture with the addition of oxygenated compound (ethanol) that improves the octane number was also determined. The studies relating to the stability was conducted for the model fuel replacing diesel. Due to the fact that during storage, middle distillates more rapidly degraded than petrol fractions, in addition to conventional ageing tests to assess the stability of the fuel at the time of storage, an accelerated ageing test at 45°C based on ASTM D 4625 [25] was conducted. The increased temperature helps to speed up the reactions occurring in the fuel during storage under conventional conditions.

2. Tests of chemical stability research of model blends

The induction period (micro method) based on the EN 16091 [24] standard was used to evaluate the chemical stability of model fuels (replacing petrol). In the micro method, induction period is defined as the time elapsed from the start of the study to the critical point determined by the pressure drop of the test system by 10% compared to the maximum pressure registered during oxidation. The fuel ageing process was performed at 140°C, the initial oxygen pressure at 500 kPa and the amount of sample at 5 ml. The principle of measurement in the above-mentioned norm is similar to the principle used in the Standard ISO 7536 [19]. The micro method is less labour intensive than the ISO 7536 standard, which normally is used to assess the stability of commercial petrol.

The chemical stability of selected petrol model mixtures was also evaluated by the ASTM D 873 method [26]. It determines the fuel potential to create resins and sediments. In this method, petrol is oxidized by 4 hours at 100°C in oxygen atmosphere at a pressure of 690–705 kPa. The test result is given as the contents of potential resins (the total amount of insoluble and soluble resins). Insoluble resins—precipitate adjacent to the glass wall of the test cell, from which the oxidized fuel was removed, precipitation and soluble resins—are determined by weight gain of test cell after the study, compared to the mass of clean test cell before the examination. The oxidation products present in the fuel after ageing in dissolved form in the oxidized fuel or as deposits adhering to the walls of the test cell, soluble in toluene-acetone, are soluble resins.

Four most commonly used methods of testing the degree of diesel oils and their surrogate degradation were selected. All of them are methods of accelerated ageing. Since one of the characteristics of the fuel ageing process is the deposit formation, two of this type of tests were selected for the research, including ASTM D 5304 [21] and EN ISO 12205 [22], where the amount of insoluble filterable sediments precipitated during the test and retained on the filter and the resin adhering to the walls of the test cell is determined. At the same time, to verify the theory of the possibility of the fuel degradation without producing deposits, it was decided on two another oxidation stability tests EN 16091 [24] and EN 15751 [23], where the result is presented in minutes or hours. The EN 15751 test is dedicated to a fuel containing at least 2% v/v of fatty acids methyl esters, and the induction period measured by this method is the time from the start of measurement to the moment, where the formation of oxidation products recorded by changes in conductivity begins rapidly increase.

The tests conditions were as follows:

ASTM D 5304—temperature, 90°C; oxygen atmosphere at a pressure of 800 kPa; time, 16 hours

EN ISO 12205—temperature, 95°C; oxygen barbotage; time, 16 hours

EN 16091 (called PetroOxy test)—temperature, 140°C; oxygen pressure of 700 kPa

EN 15751 (called Rancimat test)—temperature, 110°C; air flow

Spectrophotometric analysis of model samples or their components, before and after accelerated ageing by PetroOxy test, was carried out using a Magna 750 FT-IR spectrophotometer (Nicolet). The spectra measurements were performed in KBr cuvette having a thickness of 0.065 mm in the range of 4000 to 400 cm^{-1}.

Long-term ageing test was carried out to determine the oxidative stability of the fuel during the long-term storage. The fuel samples were placed in a thermostatic chamber at 45°C in vented glass bottles. At set intervals, another bottle with the fuel was removed from the chamber and deposit content was determined.

2.1. Model blend replacing petrol: stability test results

For the preparation of model petrol, the following hydrocarbons were used: n-heptane representing paraffins; isooctane, branched-chain alkanes; cyclohexane, cyclic alkane; toluene, aromatic compound; and cyclohexene, olefin. A non-hydrocarbon component—anhydrous ethanol—which is added to petrol to increase the octane number was also used. The organic peroxide (tertiary—butyl hydroperoxide TBHP) dosed in an amount of 50 mmol/l was used to prepare and simulate the oxidized fuel [27]. The two antioxidants, one being a mixture of sterically hindered phenols (AO-1) and the second—phenylenediamine (AO-2)—were used to improve the oxidation resistance of the selected compositions.

In addition, commercial petrol with research octane numbers 95 (95 RON) and 98 (98 RON) and fuel fractions used for petrol composing, i.e. cracked naphtha, alkylate and reformate were studied. Petrol 95 RON contained ethanol.

The composition of tested model petrol for ageing tests is shown in **Table 1**. Data concerning the major components of the commercial petrol in Europe presented in the publication [28] were used to determine the volume fractions of various hydrocarbons. The MP_2 mixture was prepared by mixing a volume ratio of 95:5 MP_1 petrol with anhydrous ethanol. Similarly, the MP_3 and MP_4 fuels were prepared, whereby cyclohexane was added instead of alcohol. The MP_5 fuel consisted of 90% v/v MP_1, 5% v/v ethyl alcohol and 5% v/v cyclohexene.

Symbol	Composition	Components (% v/v)					
		Heptane	Isooctane	Cyclohexane	Toluene	Cyclohexene	Ethanol
MP_1	–	14.5	36.5	11.7	37.3	–	–
MP_2	95% v/v MP_1 5% v/v Ethanol	13.8	34.8	11.1	35.4	–	5.0
MP_3	95%v/v MP_1 5% v/v Cyclohexene	13.8	34.8	11.1	35.4	5.0	–
MP_4	93%v/v MP_1 7% v/v Cyclohexene	13.5	34.0	10.8	34.7	7.0	–
MP_5	90%v/v MP_1 5% v/v Cyclohexene 5% v/v Ethanol	13.0	32.9	10.5	33.6	5.0	5.0

Table 1. Chemical composition of model petrol for ageing tests.

Induction period of hydrocarbons and ethyl alcohol is shown in **Table 2**. Among the tested compounds, the induction period using micro method was determined for cyclohexane, cyclohexene and ethanol. In the case of other pure compounds, after 360 min (6 hours) of test, there was no pressure drop meeting the criteria of the critical point. For comparison, induction period of fuel fractions, which are used for composing petrol, was determined. The tests were performed for three refinery fractions—the cracked fractions comprising about 20% v/v of olefins and aromatics, alkylate consisting only of paraffins and reformate—which the major component was aromatic compounds. The test results indicate that, among petroleum fractions, the cracked fraction containing unsaturated hydrocarbons undergoes the fastest oxidation.

Components	Induction period at 140°C, (EN 16091) min
Heptane	Not determined*
Isooctane	Not determined*
Cyclohexane	311.12
Toluene	Not determined*
Cyclohexene	11.31
Ethanol	298.17
Cracked naphtha	33.5
Alkylate	63.5
Reformate	79.9

*Recorded pressure drop was less than 10% after 360 min of the test.

Table 2. Induction period of pure hydrocarbons, ethanol and fuel fractions.

The values of the induction period for model mixtures determined by a micro method are shown in **Table 3**. Comparatively, the results for the commercial petrol of 95 research octane numbers (95RON) and (98RON), to which deliberately introduced 3% v/v cyclohexene (OLEF), were presented in the article. Petrol 95 RON contained 5% v/v ethanol.

	Model petrol					Commercial petrol			
	MP-1	MP-2	MP-3	MP-4	MP-5	95RON	95RON + OLEF	98RON	98RON + OLEF
Induction period, min	Not determ*	Not determ*	91.5	88.75	Not determ*	44.1	34.9	47.9	39.8

*Recorded pressure drop was less than 10% after 360 min of the test.

Table 3. Induction period determined by micro methods according to PN-EN 16091.

The MP-1 hydrocarbon fuel was chemically stable. The addition of cyclohexene, which, according to the literature [6], is easily oxidized, affects the chemical stability of the MP-1 model fuel. The higher the content of this compound, the shorter the induction period of MP-3 and MP-4 mixture. The effect of oxygen compound (ethanol) to this parameter was also checked. Ethanol did not deteriorate stability of the MP-1 fuel, moreover positively affected the stability of the fuel comprising cyclohexene—the MP-3 mixture broke after 91.5 min, and for the mixture containing both cyclohexene and ethanol (MP-5), after 6 hours (360 min) of the test decrease in oxygen pressure in the system was less than criterial 10%. In the case of commercial fuels, tested samples were characterized by a much shorter induction period than the model fuel. The addition of di-olefins (OLEF) in the amount of 3% v/v caused a sharp decline in the value of this parameter.

The obtained results of the induction period were confirmed by the tests of potential resins according to ASTM D 873, as shown in **Table 4**. As a result of the oxidation, the insoluble deposits have formed only in the case of MP-3 mixture (containing cyclo-olefin). The presence of soluble deposits has been found for all tested mixtures; the lowest value was observed for the base composition containing neither cyclohexene nor ethanol. Comparatively the oxidation of fuel samples available on the market was carried out.

Parameter	Unit	Model petrol				Commercial petrol			
		MP-1	MP-2	MP-3	MP-5	95RON	95RON + OLEF	98RON	98RON + OLEF
Soluble gum	mg/100 ml	0.1	0.3	0.3	0.2	31.9	33.9	22.3	25.5
Insoluble gum		0.0	0.0	0.4	0.0	0.2	1.1	0.3	1.2
Potential gum		0.1	0.3	0.7	0.2	32.1	35.0	22.6	26.7

Table 4. Content of potential resins in the model mixtures according to ASTM D 873.

Induction period of model petrol and the commercial petrol, where the oxidation effect was simulated by the addition of TBHP as a source of radic ' , was examined using micro method mixtures. The obtained results are shown below (**Table 5**):

Parameter	Model petrol				Commercial petrol			
	MP-1 + TBHP	MP-2 + TBHP	MP-3 + TBHP	MP-5 + TBHP	95RON	95RON + TBHP	98RON	98RON + TBHP
Induction period, min	Not determ*	Not determ*	22.3	52	44.1	17.5	47.9	26.2

*Recorded pressure drop was less than 10% after 360 min of the test.

Table 5. Induction period of model fuels and petrol with TBHP determined by the micro method.

The addition of peroxide to the model mixture resulted in a deterioration of oxidation resistance of samples containing the cyclic olefin (MP-3, MP-5). The samples without the addition of cyclohexene did not 'break' during the test run for 6 hours. In contrast, commercial fuels with intentionally introduced TBHP are characterized by significantly lower induction period when compared to fuel without the addition of this substance.

Two antioxidants were used in order to improve the oxidation resistance of the MP-3 sample. One of them, defined as AO-1, comprises a mixture of mono-*tert*-butylphenol, di-*tert*-butylphenol and tri-*tert*-butylphenol as active substance; the second (AO-2) comprises *N,N'*-di-sec-butyl-p-phenylenediamine as the active ingredient. The effect of antioxidants was evaluated by the induction period using the micro method. The activity of antioxidants in petroleum fractions was checked. The results, given in **Table 6**, clearly show that in the tested systems, the addition of an amine is more effective.

	Induction period, min				
	Without additives	+60 mg AO-1/1 kg MP-3	+100 mg AO-1/kg MP-3	+25 mg AO-2/1 litre MP-3	+43 mg AO-2/1 litre MP-3l
MP-3	91.5	Not determined*	–	Not determined*	–
Cracked naphtha	33.5	48.7	67.3	60.3	87.3
Alkylate	63.5	73.5	86.8	205	Not determined*
Reformate	79.9	114.4	149.4	Not determined*	Not determined*

*Recorded pressure drop was less than 10% after 360 min of the test.

Table 6. Induction period of MP-3 model mixture and petroleum fractions with addition of antioxidants.

The induction period of the MP-3 mixture with the addition of antioxidants, for which conditions of oxidized fuel were simulated by the addition of TBHP, was also studied. The tests were performed in two variants: (1) antioxidant and peroxide were added to the model mixture, and (2) an antioxidant was introduced to oxidized by peroxide sample. Similar results were obtained, regardless of the order of addition of the additive and peroxide. **Table 7** contains the induction period for option 2.

	Model fuel MP-3				
	+ TBHP	+ TBHP+60 mg/kg AO-1	+ TBHP+100 mg/kg AO-1	+ TBHP+25 mg/l AO-2	+ TBHP+43 mg/l AO-2
Induction period, min	22.3	22.5	23.5	32.1	36.0

Table 7. Induction period of MP-3 model mixture containing antioxidant additive and TBHP.

Antioxidants, both phenolic and amine, advantageously improved chemical stability of the fuel. However, their effectiveness was influenced by the presence of superoxide radicals, which is confirmed by the test results shown in the table above. It can be assumed that antioxidants introduced into the fuel, which is already occurring processes associated with oxidation, do not fulfil their protective function.

2.2. Model blend replacing diesel oil: stability test results

Diesel fuel, same as petrol, is a mixture of many hydrocarbons of various structural features. The following hydrocarbon groups are in the oil: n-alkanes, iso-alkanes, cycloalkanes, olefins, aromatics and polyaromatics. Most of hydrocarbons have 15–17 carbon atoms in the molecule, the least hydrocarbons containing less than 9 and more than 23 carbon atoms in the molecule. Diesel fuel usually contains 40–50% by weight of paraffins, 20–30% by weight of cycloparaffins, 5–15% by weights of aromatics, 2–7% by weight of olefins and up to 7% by weight of fatty acid methyl esters (FAME). A disadvantage of esters is their low resistance to oxidation. In contrast to the 'pure' diesel fuel, esters added to the fuel make it more susceptible to oxidation or autoxidation during prolonged storage.

In the present study, n-hexadecane (n-cetane), n-butyl cyclohexane, n-butyl benzene and fatty acid methyl esters were used to compose the model diesel fuel.

The first stage of the study included accelerated oxidation tests of hydrocarbons and fatty acid methyl esters. An induction period examination (according to EN 16091 standard) has been used for this purpose. The analysis of possible changes in the quality of oxidized samples of mono-compounds on the basis of the IR spectrum was also conducted. The induction period results are shown in **Table 8**.

Component	Induction period at 140°C (EN 16091) min
n-Cetane	80.57
n-Butyl cyclohexane	61.83
n-Butyl benzene	175.55
FAME	27.33

Table 8. Induction period of pure ingredients and FAME.

For all the tested substances, there was a 10% decrease in oxygen pressure in the measuring chamber (on this basis the induction period was determined). N-butyl cyclohexane (saturated cyclic hydrocarbon) was the least resistant to oxidation at elevated temperature. N-butyl benzene (aromatic hydrocarbon with an aliphatic side chain) was the most stable compound. Among the four tested components, fatty acid methyl esters demonstrated the lowest resistance to oxidation. The changes that occurred as a result of accelerated oxidation of each model hydrocarbons were visible in IR spectrum. Products of hydrocarbons oxidation occurred in two spectral ranges. For a hydroxyl group, a broad band was from 3600 to 3300 cm^{-1} (with a maximum of about 3500 cm^{-1}), whereas for the carbonyl group, bands were in the range from 1750 to 1650 cm^{-1} (characteristic for oxidation products such as carboxylic acids, aldehydes, esters, ketones). Each position of the bands may undergo some shifts depending on the structure of the compound and the proximity of other groups. **Figure 1** shows an example of one of the tested hydrocarbon (n-cetane) spectra, before and after accelerated oxidation.

In the spectrum of n-cetane after oxidation, there are three peaks at a wavelength of 1710 cm^{-1}, 1714 cm^{-1} (area A) and 3558 cm^{-1} (area B). These peaks are not present in the spectrum of the n-cetane before oxidation. Peaks at 1710 cm^{-1} and 1714 cm^{-1} are characteristic for carbonyl groups (C=O) and are evidence of carbonyl compound formation in the n-cetane, while the third band at 3558 cm^{-1} indicates the formation of hydroxy compounds (—OH). A similar spec-

http://dx.doi.org/10.5772/67056

trum was obtained for the oxidized n-butyl cyclohexane, with the difference that was only one characteristic carbonyl peak. In the spectrum of the n-butyl benzene, changes were observed only in a range of the carbonyl band, and compounds with hydroxyl group were not formed. In the spectra of fatty acid methyl esters, there were no changes within the carbonyl band (the band is characteristic for esters), whereas area under the hydroxyl peak significantly increased.

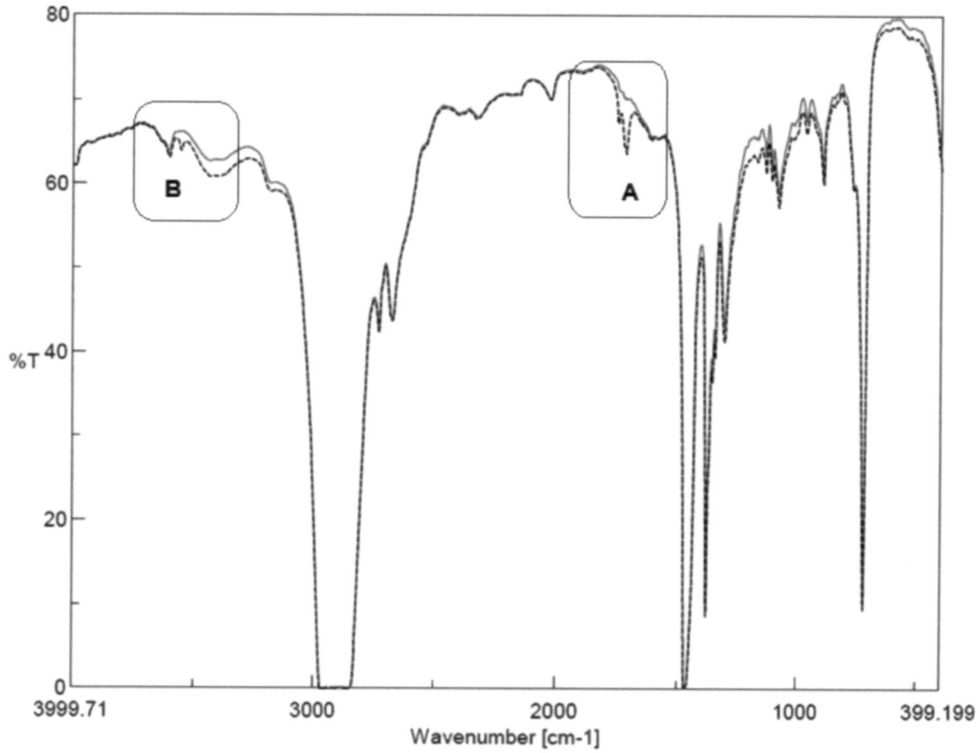

Figure 1. Comparison of the n-cetane before oxidation IR spectra (solid line) and after oxidation (dashed line).

The model mixtures (the composition is given in **Table 9**) were prepared for the tests of resistance to oxidation. Two mixtures were selected: the first mixture composed exclusively of hydrocarbons (M0) and the second one composed of hydrocarbons enriched with biocomponent—fatty acid methyl esters (M7).

Components	**Model fuel M0, %(v/v)**	**Model fuel M7, %(v/v)**
n-Cetane	50	46.5
n-Butyl cyclohexane	28	26.0
n-Butyl benzene	22	20.5
FAME	0	7.0

Table 9. Composition of the model blends.

The test of accelerated oxidation according to the previously described EN 16091 standard (PetroOxy test) was used for the mixture oxidation. In the case of the M7 mixture (containing FAME), additionally the oxidative stability examination according to EN 15751 (Rancimat test) was performed. The results are shown in **Table 10**.

Model fuel	Oxidative stability (EN 15751), h	Induction period (EN 16091), min
M0	–	110.90
M7	38.20	92.58

Table 10. Results of the oxidative stability of the blends of M0 and M7.

For the both studied mixtures, there was a 10% decrease in oxygen pressure in the measuring chamber; therefore, the induction period has been determined. The M7 mixture, due to the presence of methyl esters, was less resistant to oxidation than M0 blend. Comparative analysis of the IR spectra showed that in the M0 mixture after oxidation carbonyl compounds with characteristic bands 1715 cm^{-1} and 1743 cm^{-1}, shown in **Figure 2** within D area and wide, expanding bands characteristic for hydroxyl groups (C area). In the spectrum of oxidized M7 mixture containing fatty acids methyl esters are not substantially changed within the carbonyl band, because in this spectral range lies peak characteristic for esters. There was however a slight increase of the area of 1743 cm^{-1} the carbonyl band with a typical inflexion point of the peak at 1743 cm^{-1}, which indicates the formation of oxidation products. In **Figure 2** the spectra of the M0 mixture before and after oxidation are shown.

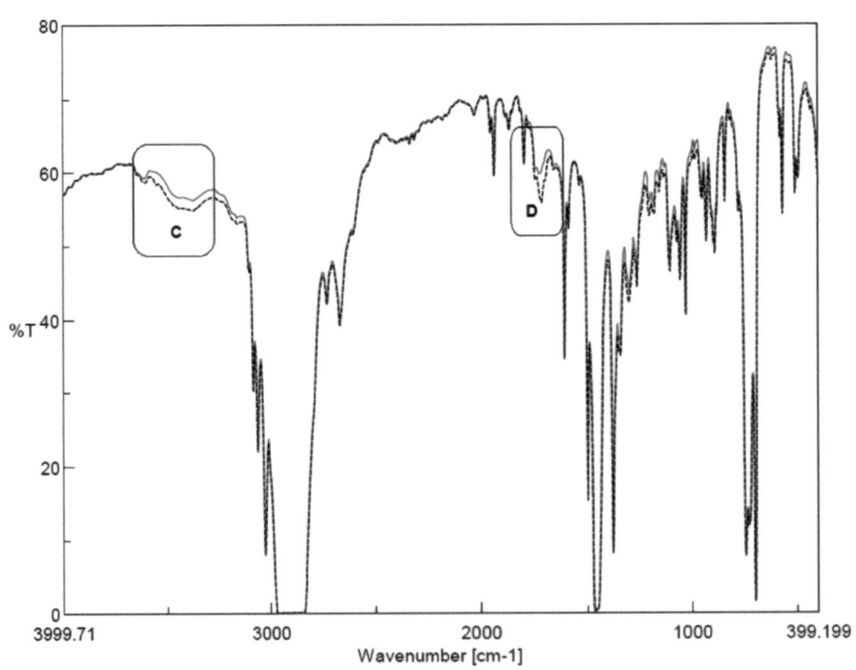

Figure 2. Comparison of the M0 mixture before oxidation (solid line) and after oxidation (dashed line) IR spectra.

2.2.1. *Long-term ageing test*

2.2.1.1. *Model diesel oil*

The M0 and M7 model mixtures were stored for 6 months. Every 30 days, measurements of oxidation stability (according to EN 16091 and EN 15751) have been performed, and the amount of sediment has been determined (according to ASTM D5304) for the batch of stored mixtures. The results are shown in **Table 11**.

Storage time, day	Model fuel M0		Model fuel M7		
	Induction period EN 16091, min	Deposit content ASTM D 5304, mg/100 ml	Induction period EN 16091, min	Oxidation stability EN 15751, hr	Deposit content ASTM D5304, mg/100 ml
0	110.90	0.6	92.58	38.2	0.9
30	109.45	0.4	91.90	38.0	0.7
60	110.25	0.6	91.25	37.7	1.1
90	108.83	0.8	90.84	37.8	1.5
120	108.30	0.7	88.55	37.3	2.8
150	107.15	1.0	88.05	37.1	2.6
180	105.52	0.9	86.62	36.4	3.5

Table 11. Results of model fuels storage stability test at elevated temperature.

The M0 and M7 model mixtures had a high stability during conducting long-term ageing test. Induction period measured by PetroOxy test varied slightly for both samples M0 and M7. At the end of the test, threefold increase of the total sediment for the M7 sample was observed. The M0 mixture (without FAME) contained the amount of sludge at a similar level. The M7 mixture had high oxidation stability, 38.2 hours, which decreased slightly during 180 days of storage to 36.4 hours.

2.2.1.2. *Diesel oil with various content of FAME*

Two samples of commercial diesel fuel containing various amounts of fatty acid methyl esters were examined. The sample A contained 1% by weight of FAME, while sample B, 7% by weight of FAME. The samples were stored at 45°C for 6 months in a dark place and at 30-day intervals were collected for testing in order to determine the amount of deposits formed during fuel storage (ASTM D 5304 test) and determine the induction period (EN 16091 test, EN 15751 test). The results are shown in **Table 12**.

In addition to the above-mentioned tests, the oxidation stability test (according to EN ISO 12205) was performed. The test was performed at the beginning and after 180 days of storage the samples A and B.

At the beginning of the study of the A sample, 7 g/m^3 total sludge including 4 g/m^3 filterable sludge and 3 g/m^3 resins were produced while for the same sample after 6 months of storage, 25 g/m^3 filterable sludge and 2 g/m^3 resins, 27 g/m^3 total.

Storage time, day	Sample A (diesel fuel with 1% FAME)		Sample B (diesel fuel with 7% FAME)		
	Induction period EN 16091, min	Deposit content ASTM D 5304, mg/100ml	Induction period EN 16091, min	Oxidation stability EN 15751, hr	Deposit content ASTM D5304, mg/100ml
0	41.9	1.4	31.1	27.5	1.9
30	38.8	2.4	29.9	27.8	3.7
60	39.7	3.7	25.5	26.4	3.5
90	40.7	4.4	27.8	27.9	11.5
120	37.5	8.2	26.9	27.7	12.8
150	39.5	5.7	26.7	27.7	8.6
180	32.8	7.7	26.6	27.5	9.5

Table 12. Results of commercial diesel oil storage stability test at elevated temperature.

The same for the B sample, at the beginning of the test, the amount of filterable sludge was 7 g/m^3, and 1 g/m^3 resins, 8 g/m^3 total while for the same sample after 6 months of storage, 56.2 g/m^3 filterable sludge and 2 g/m^3 resins, 58.2 g/m^3 total. The presented data show that with the increase of the sample storage period, the total amount of deposits increases several times. The increase of sediments was mainly caused by the formation of filterable deposits, while the amount of resins did not change.

After 6 months of fuel storage, induction period determined by EN 16091 was lower than at the beginning of the study, for both samples A and B. For A sample the induction period dropped from 41.9 to 32.8 min and for the B sample from 31.1 to 26.6 min. The induction period measured by Rancimat test for B sample containing 7.2% FAME was constant during 180 days of storage fuel.

3. Conclusions

The model fuel as a petrol substitute composed of four hydrocarbons (heptane, isooctane, cyclohexane, toluene) was chemically stable at elevated temperature and oxygen pressure. The addition of a reactive olefin, i.e. cyclohexene caused a noticeable reduction in oxidation resistance determined by an induction period—the higher the content of this compound in the fuel, the lower the chemical resistance. The results of the induction period were confirmed by the study of potential resin content—as a result of oxidation, most deposits were formed in fuels with cyclohexene. Ethanol introduced in an amount of 5% v/v to the model fuel does not have a negative effect on the induction period, whereas the fuel with a cyclic olefin improved the value of this parameter. Antioxidants (phenolic and amine) added to the fuel with cyclohexene improved effectively oxidation resistance. The test of induction period of model fuels, where oxidation was simulated by the addition of TBHP as a source of reactive radicals, showed deterioration of the parameter sample cyclohexene.

Comparative studies conducted using commercial petrol have shown that the model fuel reflects behaviour of real fuel to a limited extent. The main drawback is the lack of a 'breaking' of MP-1 model fuel designating the induction period of accelerated oxidation tests in micro-scale. It can be assumed that in the model, fuel lacked representative olefin susceptible to oxidizing agents but less reactive than cyclohexane.

The diesel fuel was replaced by the model mixtures, which were prepared from three to four compounds, representative for hydrocarbon groups presented in commercial fuel. Applied hydrocarbons are characterized by a high resistance to oxidation (the induction period was over 60 min). Fatty acid methyl esters, which have been used in studies as components from renewable sources, were the least resistant to oxidation. The IR spectrophotometric analysis of individual hydrocarbons and model mixtures showed that oxidized products with carbonyl or hydroxyl binding (carboxylic acids, aldehydes, ketones) were formed as a result of oxidation.

During long-term ageing tests at elevated temperature, the M0 and M7 model mixtures are characterized by high stability, and the induction period varied slightly with time. Only for the M7 mixture was an increase in the amount of deposits after 6 months of storage, which probably was related with the presence of fatty acid methyl esters in the studied system.

In the case of long-term ageing test of commercial A and B fuels, an increase in the amount of deposits according to ASTM D5304 after 4 months of storage was observed. In contrast, the oxidative stability study using EN ISO 12205 method showed a fourfold and sevenfold increase in amount of deposits, respectively, for A and B samples, indicating that the degradation proceeds more quickly in the fuels with the higher FAME. The oxidation stability of A and B samples decreased with increasing time of storage.

Based on the obtained results, it can be concluded that the model blends of diesel fuel of the composition shown in **Table 9** may be used to predict the oxidative stability of diesel fuels using the method of the induction period EN 16091, and for the samples containing fatty acid methyl esters, the EN 15751 method can also be used. However, model mixtures cannot be used to predict the rate of degradation of diesel fuel expressed as the amount of filterable sediments and resins. The model mixtures, during long-term ageing process, generate a small amount of sediments constant at the same level.

Author details

Małgorzata Odziemkowska, Joanna Czarnocka* and Katarzyna Wawryniuk

*Address all correspondence to: j.czarnocka@pimot.eu

Automotive Industry Institute, Liquid Fuels and Bio-Economy Department, Warsaw, Poland

References

[1] Czarnocka J., Matuszewska A., Odziemkowska M. Autoxidation of Fuels During Storage. In: Biernat K., editor. Storage Stability of Fuels. InTech; 2015. pp. 157–188. DOI: 10.5772/5979 Available from: http://dx.doi.org/10.5772/59807

[2] Batts B.D. Zuhdan F.A. A Literature Review on Fuel, Stability Studies with Particulate Emphasis on Diesel Oil. Energy Fuels 1991;5(1):2–21

[3] Pedersen C.J.. Mechanism of Antioxidant Action in Gasoline. Industrial Engineering and Chemistry 1956;48(10):1881–1884

[4] Pedersen C.J. Inhibition of Deterioration of Cracked Gasoline During Storage. Industrial Engineering and Chemistry 1949;41(5):924–928

[5] Nagpal J.M., Joshi G.C., Rastogi S.N. Stability of Cracked Naphtha from Thermal and Catalytic Processes and their Additive Response. Part I. Evaluation of Stability and Additive Response. Fuel 1995;74(5):714–719

[6] Pereira R.C.C., Pasa V.M.D. Effect of Mono-Olefins and Diolefins on the Stability of Automotive Gasoline. Fuel 85;(2006):1860–1865

[7] Pereira R.C.C., Pasa V.M.D. Effect of Alcohol and Copper Content on the Stability of Automotive Gasoline. Energy Fuel 2005;19:426–432

[8] Marshman S.J., Davia P. Storage Stability of Distillate Diesel Fuels: Changes in Phenalene and Phenalenone Concentrations During Long Term Ambient Storage. Preprint Paper American Chemical Society Division of Fuel Chemistry 1990;35:41108–41116

[9] Beranek L.A., McVea G.G., O'Connell M.G., Solly R.K. Rates of Indole—Phenalenone Reactions in Middle Distillate Fuel. Preprint Paper American Chemical Society Division of Fuel Chemistry 1990;35 4:1117–1124

[10] Li D. et al., Spectroscopic Studies on Thermal-Oxidation Stability of Hydrocarbon Fuels. Fuel (2008);87:3286–3291

[11] Zabarnick S.. Chemical Kinetic Modeling of Jet Fuel Autoxidation and Antioxidant Chemistry. Industrial and Engineering Chemistry Research 1993;32:1012–1017

[12] Karavalakis G., Hilari D., Givalou l., Karonis D., Stournas S. Storage Stability and Ageing Effect of Biodiesel Blends Treated with Different Antioxidants. Energy (2011);36:369–374

[13] Karavalakis G., Stournas S., Karonis D. Evaluation of the Oxidation Stability of Diesel/ Biodiesel Blends. Fuel (2010);89:2483–2489

[14] Wachal A., Propellants and motor oils for contemporary piston-, jet- and rocket- engines, published by Ministry of National Defence (Poland), 1959

[15] Denisov E.T., Afanas'ev I.B. Oxidation and Antioxidants in Organic Chemistry and Biology. CRC Press, Taylor & Francis Group, Abingdon UK; 2005.

[16] Altin O., Ester S. Carbon Deposit Formation from Thermal Stressing of Petroleum Fuels. Preprint Paper American Chemical Society Division of Fuel Chemistry. 2004;49(2):764

[17] Gernigon S., Sicard M., Ser F., Bozon-Verduraz F. Hydrocarbon Liquid Fuels Thermal Stability, Antioxidants Influence and Behaviour. Proceedings of 11th International Conference on Stability, handling and Use of Liquid Fuel 2009, Prague, Czech Republic, 18–22 October 2009, pp. 472–509

[18] Natelson R.H., et al. Experimental Investigation of Surrogates for Jet and Diesel Fuels. Fuel (2008);87:2339–2342

[19] Standard ISO 7536. Petroleum Products—Determination of Oxidation Stability of Gasoline—Induction Period Method

[20] Standard ISO 6246. Petroleum Products—Gum Content of Light and Middle Distillate—Jet Evaporation Method

[21] ASTM D 5304 Standard Test Method for Assessing Middle Distillate Fuel Storage Stability by Oxygen Overpressure

[22] Standard EN—ISO 12205 Petroleum Products — Determination of the Oxidation Stability of Middle-Distillate Fuels

[23] Standard EN 15751 Automotive Fuels—Fatty Acid Methyl ester (FAME) Fuel and Blends with Diesel Fuel—Determination of Oxidation Stability by Accelerated Oxidation Method

[24] Standard EN 16091 Liquid Petroleum Products—Middle Distillates and Fatty Acid Methyl Ester (FAME) Fuels and Blends—Determination of Oxidation Stability by Rapid Small Scale Oxidation Method

[25] ASTM D 4625 Standard Test Method for Middle Distillate Fuel Storage Stability at 43°C (110°F)

[26] ASTM D 873 Standard Test Method for Oxidation Stability of Aviation Fuels (Potential Residue Method)

[27] Harrigan Sr. M., Banda A., Bonazza B., Graham P., Slimp B. A Rational Approach to Qualifying Materials for Use in Fuel Systems. In: SAE Technical Paper Series 2001-01-2013, ISNN 0148–7191

[28] Pera C., Knop V. Mint: Methodology to Define Petrol Surrogates Dedicated to Auto-Ignition in Engines. Fuel 2012;96:59–69

Chapter 2

Effect of Waste Cooking Oil Biodiesel Blends on Performance and Emissions from a CRDI Diesel Engine

Giancarlo Chiatti, Ornella Chiavola and Erasmo Recco

Additional information is available at the end of the chapter

http://dx.doi.org/10.5772/intechopen.69740

Abstract

The employment of biofuels in blends with diesel oil proved to attain a reduced environmental impact without compromising the engine performance. Among biofuels, waste cooking oil offers the advantages of its reduced raw material cost in comparison with fresh vegetable oil cost; it also eliminates the environmental impacts caused by its disposal. Although a great number of researches has been devoted to biodiesel combustion in engines and pollutant emissions, few studies can be found on light duty diesel engine equipped with up-to-date technologies. This work aims at investigating the impact of waste cooking oil percentage in blends with diesel oil on the performance and emission characteristics of an up-to-date light and compact common rail diesel engine whose main application is in microcars and in urban vehicles. A comprehensive experimental activity was performed in the engine complete operative field. The comparison of the results with those obtained with standard ultralow-sulfur diesel highlighted that the engine performance was quite similar for B20 and diesel oil. B40 suffered for the lower caloric value in regard to diesel. A reduction in CO and HC was obtained with biodiesel blends, along with an increase in NO_x. Particulate emissions were also reduced for biodiesel blends; the mean size of particles was smaller as regards diesel oil.

Keywords: diesel engine, biodiesel blend, waste cooking oil, pollutant emission, particulate emission

1. Introduction

In the frame of complying with the emission regulations that become day by day more and more stringent, researchers have focused their interest on areas of fuel injection control strategies, exhaust gas recirculation, exhaust gas posttreatment devices, and also on areas of alternative fuels.

Alternative fuels from vegetable oils and animal fats have been proposed for a partial and total replacement of diesel fuel to reduce the environmental impact in terms of air pollution and dependence on fossil fuel.

Among these fuels, biodiesel from vegetable oils has received great attention for its renewability and its potential to reduce greenhouse gas emissions and soot formation [1–4].

Experimental investigations have highlighted that biodiesel used in blends with diesel is responsible for a reduction in unburned hydrocarbon, carbon monoxide, and particle emissions due to the increased oxygen content in the fuel [4–8]. In regard to NO_x emission, somehow contradictory conclusions were found, since there are numerous factors, each has its own relative importance according to the engine technology and operating conditions of the blended fuel [9–13]. Physical properties, chemical composition, and structure of the biodiesel alter the fuel injection and ignition process, and then the combustion development and the engine exhaust emissions [14–18]. Many studies proved that biodiesel feedstock and blend ratios have a large impact on obtained results. Peng [2] tested various types of biodiesel on a turbocharged diesel engine; he found smoke opacity, CO and HC decreased, but fuel consumption increased compared to petrol diesel. Serrano et al. [11] analyzed the behavior of an EURO 5 engine fuelled with two biodiesel blends (7 and 20% v/v). Fuel consumption was not consistently increased with biodiesel; NO_x emission with biodiesel use did not present significant rise. Yehliu et al. [19] investigated the impact of fuel properties and injection strategy on the combustion process and soot emission. Three fuels were tested on a turbocharged diesel engine, and particle size distribution was measured. Ajtai et al. [20] studied the effect of fuel type and engine condition on number and size distribution of diesel soot. They found that the biodiesel content in the total fuel amount can modify the characteristics of the exhaust particles.

Among all suitable biodiesel fuels, waste cooking oil (WCO) has been considered a promising alternative to vegetable fresh oil because of its reduced raw material cost (the price of WCO is two to three times cheaper than virgin vegetable oils [21]). Moreover, WCO conversion into fuel offers the advantage of eliminating the environmental impact caused by its disposal. Previous studies demonstrated the suitability of WCO as a biofuel. Attia and Hassaneen [12] studied the effect of various WCO blends on the performance of a single-cylinder diesel engine. The best value of a brake specific fuel consumption was attained at blended fuel containing 20% of WCO. A range of blending ratio between 20 and 50% v/v showed the best environmental behavior. Gopal et al. [22] investigated the performance and emission characteristics of a single-cylinder diesel engine designed for agricultural purpose fuelled with WCO and its blends. The study revealed that WCO has lower CO, HC, and smoke opacity than diesel. On the other hand, NO_x and specific fuel consumption were higher than diesel. An et al. [23, 24] evaluated the influence of WCO biodiesel/blends on combustion and exhaust emission characteristics of a four-cylinder turbocharged diesel engine. The results showed that the use of WCO resulted in higher brake specific fuel consumption with regard to the exhaust emissions; the use of WCO generally resulted in a lower CO_2 and HC; under most of the operating conditions, NO_x produced by biodiesel (100% v/v) was lower compared to that of diesel fuel. Can [25] experimentally investigated the combustion development of a single-cylinder diesel engine fuelled with WCO blended in 5 and 10% with diesel fuel. An increase in NO_x emissions and a decrease in smoke and HC was found. Cheung et al. [26] analyzed the impact on the emissions from a diesel engine at fixed engine speed fuelled with diesel blended with different proportions of biodiesel from WCO. The results showed that biodiesel leads to a reduction in HC, CO, and particulate mass concentration and number concentrations, but an increase in NO_x. Man et al. [27] studied the effect of diesel operating conditions on the particle size emitted by a diesel engine operating with WCO biodiesel. They found more particles with larger size at lower engine speed; primary particles tend to form at lower engine load. Hwang et al. [28] investigated the combustion and emission characteristics of WCO and conventional diesel fuel in an optically accessible diesel engine. WCO had the benefits in CO, HC, and PM reduction at low load. In the high engine load, the emission characteristic of WCO was deteriorated than that of diesel.

Although the research on biodiesel combustion and emissions is considerable and many studies are related to the usage of WCO in diesel engines, they mainly focused on multicylinder-diesel engine of large displacement. Only some works are devoted to light-duty diesel engines, designed for agricultural purpose, and the results are mainly related to a fixed value of engine speed and load.

The main aim of this research is to analyze in detail the impact of WCO used in different percentages with diesel oil in an up-to-date light and compact, common rail diesel engine whose main application is in microcars and in urban vehicles. A comprehensive experimental activity was performed in the engine complete operative field in order to characterize the engine performance and emissions. NO_x, CO, CO_2, HC, and soot concentration were analyzed. The influence of WCO content in the blend on the particle emissions was also investigated, in terms of soot particles' size distributions.

2. Apparatus and tests

2.1. Experimental setup

A common-rail water-cooled two-cylinder diesel engine was tested in this study. Its main technical data are presented in **Table 1**. The engine was connected to an asynchronous motor (Siemens 1PH7, nominal torque 360 Nm, power 70 kW) and was installed in the test bed of the Engineering Department at Roma Tre University.

Torque measurement was carried out by means of HBM T12 (it is a strain gauge transducer with an optical encoder).

AVL Fuel Balance 733 was used for fuel consumption measurement.

The in-cylinder pressure was measured with a piezoelectric transducer AVL GU13P.

Engine type	Common rail, naturally aspirated, water-cooled
Cylinders	2
Displacement	440 cm^3
Bore	68 mm
Stroke	60.6 mm
Compression ratio	20:1
Maximum power	6.7 kW @ 3600 rpm
Maximum torque	20 Nm @ 2400 rpm

Table 1. Engine specifications.

The engine exhaust emissions (CO, CO_2, HC, O_2, and NO_x expressed as NO equivalent) were measured with Bosch BEA352. AVL particle counter (APC) and AVL micro soot sensor were used to measure the nonvolatile particle number concentration in the size range 23 nm–2.5 μm and the soot concentration in the engine exhaust gas, respectively. Particulate matter size was measured through Cambustion DMS500. This device uses a classifier column to compute the particle size distribution in the range 5 nm–1 μm, with a size resolution of 16 channels per decade. Exhaust gas passes first through a cyclone separator in order to remove particles above the measurement range (1 μm). Two stages of dilution are applied before the sample gas passes through a corona charger and into the classifier column. Primary and second dilution rates were set to 5:1 and 400:1, respectively. The charged particles flow within a particle-free sheath flow and are deflected toward grounded electrometer rings by their repulsion from a central high voltage rod. Their landing position is a function of their charge and their aerodynamic drag. Further details may be found in Ref. [29].

The engine speed was measured using an angular sensor (AVL 364C) with 2880 pulses/revolution.

Figure 1 shows the complete engine setup.

The sampling frequency was varied according to the engine speed in order to ensure a fixed angular resolution of the signals.

Data acquisition was controlled by means of LabVIEW software, by using a custom program [30].

2.2. Fuel

A second-generation biodiesel was used in the experimentation. It was obtained starting from a mixture of waste cooking oils. Due to its poor quality, it required some treatments in order to become similar to a product obtained from refined vegetable oils. A first-stage self-cleaning disk separator was used to remove 90% of the water containing the water-soluble matter and solids; a second-stage disk separator machine was used to remove the left over water. Physical deacidification was also needed to remove free fatty acids (FFA) due to product deterioration as a consequence of the use in food cooking. The neutralized products were then converted via

Figure 1. Engine setup.

a transesterification process. The resulting raw biodiesel, coming from poor raw material, was distilled in order to comply with the reference specifications of biodiesel (EN 14214). Details of the procedure may be found in Ref. [31]. The properties of the biodiesel and ultralow-sulfur diesel (ULSD) are listed in **Table 2**. The chemical composition of WCO is reported in **Table 3**.

2.3. Experiments

Since the aim of this work was to investigate the potential use of biodiesel blends in a small displacement diesel engine, some preliminary investigations were performed with the objective of selecting the maximum percentage of WCO in the blend that was able to be tested without the need of modification to the engine hardware. It was established that 40% of WCO in the blend was the highest quantity of biodiesel that could be tested, since higher percentage of biodiesel in the blend caused a degradation of the rubber hoses/seals in the engine fuel system. The experimentation was thus performed with a biodiesel percentage lower than 40% by volume: standard ultralow-sulfur diesel, B20 (80% ULSD and 20% biodiesel, by volume), and B40 (60% ULSD and 40% biodiesel, by volume) were tested. This allowed the investigated blends to be ready for use in actual engine. Before each new fuel was tested, sufficient time was given to the engine to consume the remaining fuel in the supply system.

Engine speed was varied in the engine complete operative range (2400–3600 rpm).

Load condition was varied in the field 50–80% as regards the available torque at full-load condition evaluated using diesel fuel. The maximum value was established by testing the engine with the different blends and by computing the load that are able to ensure the same value for all the tested fuels.

	ULSD	**Biodiesel**
Density [kg/m^3 at 15°C]	830	877
Viscosity [cSt at 40°C]	2.5	4.4
Lower heating value [MJ/kg]	43.1	37.1
Cetane number	52	56

Table 2. Biodiesel and ULSD fuel properties.

Mass fraction	Biodiesel
Carbon	0.812
Hydrogen	0.065
Oxygen	0.117
Sulfur	0.006

Table 3. Biodiesel composition.

The data acquisition started after the engine warm-up in order to let the engine reach nominally stationary conditions. For each running condition, 25 engine cycles were used to average the signal, thus to attenuate the engine cycle irregularities (the increase in this number did not change the feature of the trends).

3. Results

Experimental results obtained with WCO blends are shown and discussed with the aim of highlighting the characteristics of performance (torque, brake specific fuel consumption, brake thermal efficiency) and pollutant emissions (NO_x, CO, CO_2, HC, soot concentration, particle number concentration, particle size distributions).

Figure 2 shows the variation of engine torque with speed at full-load condition for diesel fuel, B20, and B40. The torque trend at full-load condition depends on the percentage of biodiesel in the fuel; since WCO has a lowering heating value than ULSD, the engine torque values related to B40 are the lowest at all engine speeds.

In order to allow the comparison between data obtained for the different fuels, it was established to perform tests at 80% of full load evaluated using diesel fuel, so as to impose the same value of load to the engine for all tested fuels.

Figure 3 shows the variation of brake specific fuel consumption (BSFC) with engine speed for 100% load. The fuel consumption increases with the content of WCO in the fuel. This is to attribute to the reduction in energy content in the biodiesel as regards diesel fuel. The average increase in BSFC over all engine speed values is 3.9% for B20 and 7.1% for B40.

Brake thermal efficiency (BTE) versus engine speed is plotted in **Figure 4**. It was evaluated by computing the ratio of the brake power to the power provided by the consumed fuel at full-load condition. The differences in the B20 and B40 averaged values are only about 1% as compared to diesel fuel.

Exhaust temperature is a very important indicator of the combustion process and has a key role in the formation of pollutants. **Figure 5** presents the variation of exhaust temperature with engine speed obtained with ULSD, B20 and B40 at 80% of load. The thermocouple was placed just downstream junction of the two-branches that connects the cylinders to the exhaust duct. All fuels are characterized by an increase of temperature with engine speed. The trends show a reduction in exhaust temperature with the increase in biodiesel ratio in the blend. This behavior

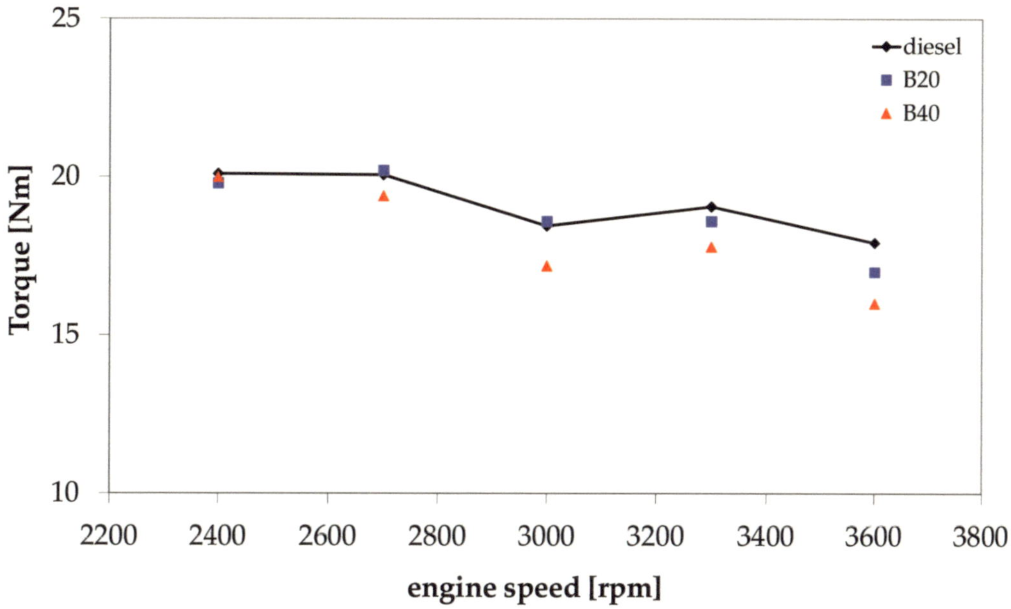

Figure 2. Variation of engine torque with engine speed [31].

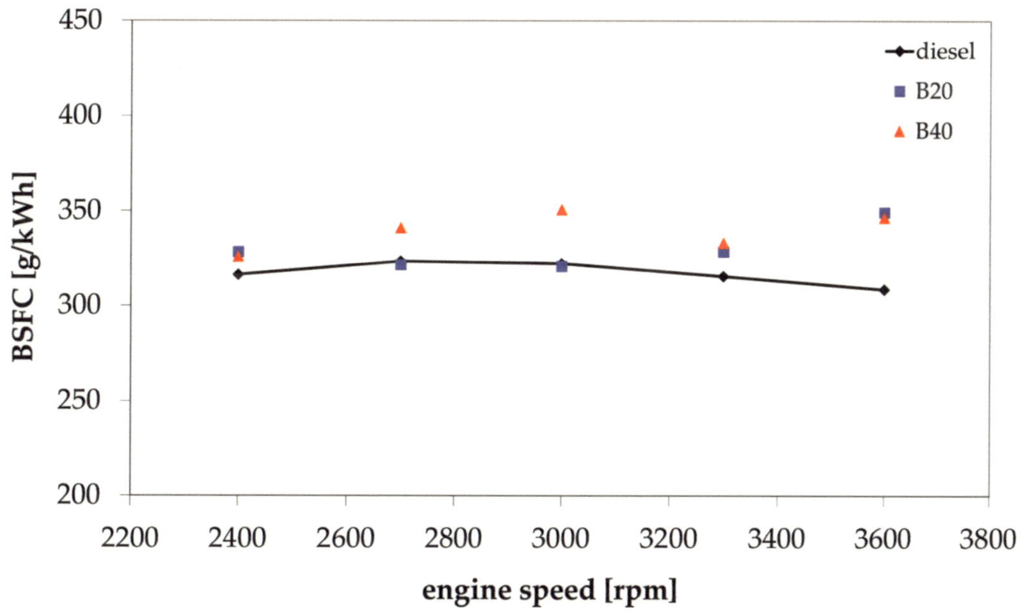

Figure 3. Variation of brake specific fuel consumption with engine speed [31].

is ascribed to the lower heating value of biodiesel, which reduces the amount of total energy released, thus reducing the combustion peak temperature and then the exhaust temperature. Data from literature are contradictory: some authors report that biodiesel has a higher combustion temperature as regards diesel fuel [13]; other authors assert the opposite behavior [23].

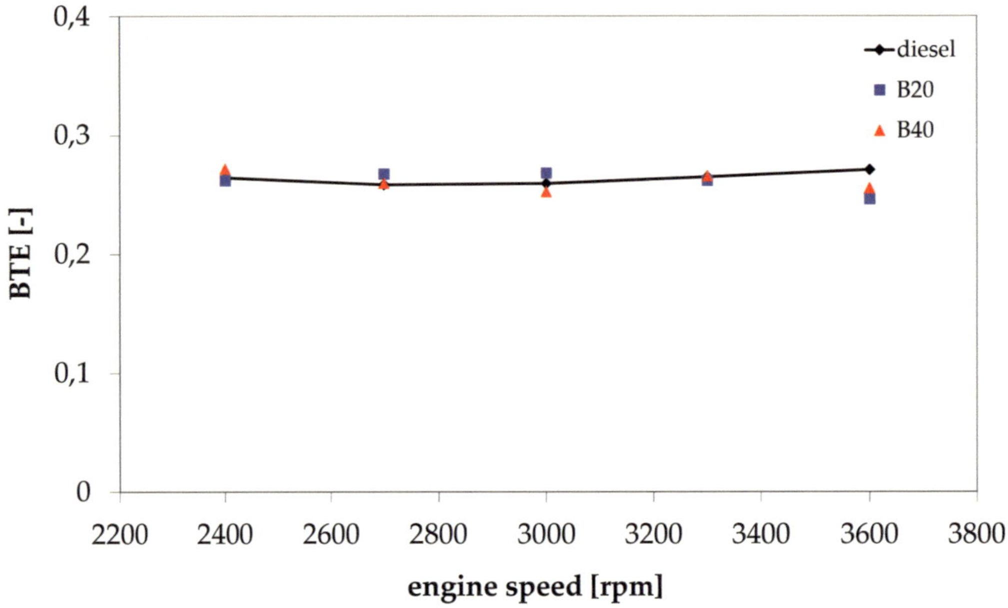

Figure 4. Variation of brake thermal efficiency with engine speed [31].

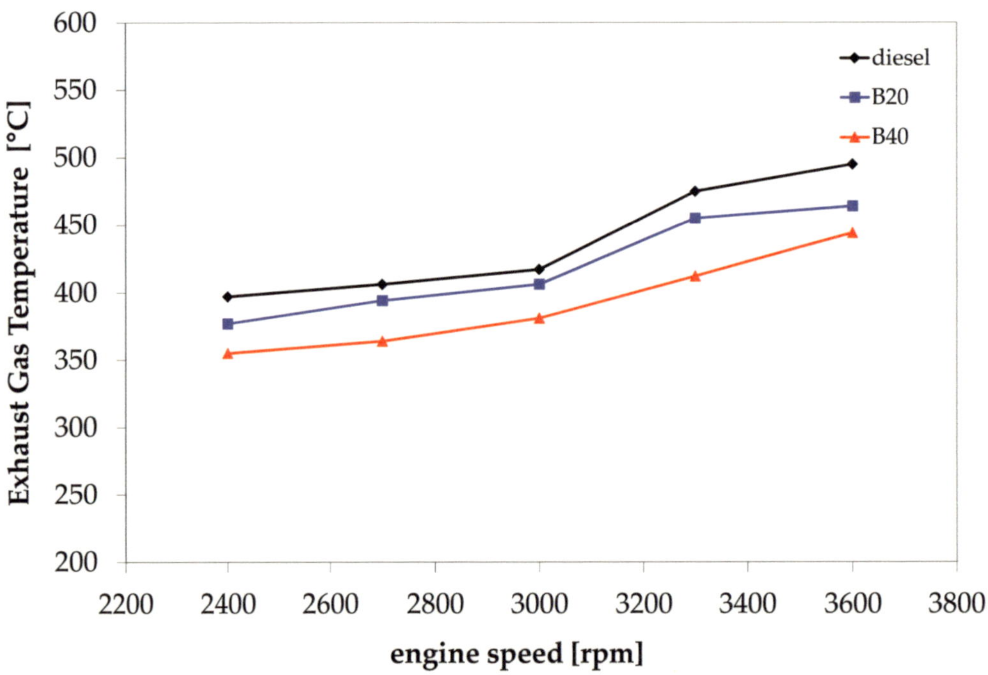

Figure 5. Variation of exhaust temperature with engine speed [31].

Figure 6 shows the variation of NO_x emission with engine speed at a fixed value of load condition (80%). It is expressed as NO equivalent. All fuels exhibit a decrease in NO_x with the increase in engine speed, in agreement with literature [32]. This trend is due to the increase in the gas motion in the cylinder at higher engine speed that is responsible for a faster mixing between fuel and air and a shorter ignition delay. At higher engine speed, the residence time of high temperature within the cylinder is shortened and this causes a reduction in NO_x emission, in spite of the temperature trends shown in **Figure 5**. The traces in **Figure 6** do not exhibit remarkable differences; B40 shows the greatest increase in NO_x, which is evident for values greater than 3000 rpm. Such increase in NO_x emission can be explained by considering the effect of temperature, the differences in fuel chemistry, spray properties, and ignition delay that affect the duration of premixed and diffusion burn regimes [4, 9, 19].

In **Figure 7**, CO emission trends obtained at 80% load are shown. The oxygen content in the biodiesel blends enhances the mixing process between air and fuel, thus allowing a reduction in CO emissions for B20 and B40 as regards diesel fuel, in agreement with results from literature [7, 33].

Figure 8 presents the variation of HC emission with respect to the engine speed. The WCO content in the blend causes a reduction in emissions as regards diesel fuel, according to published data [7, 18].

CO_2 emission trends are shown in **Figure 9**. Similar behavior has been obtained for all tested fuels. The literature reports contradictory results in this field. Some authors [34] obtained

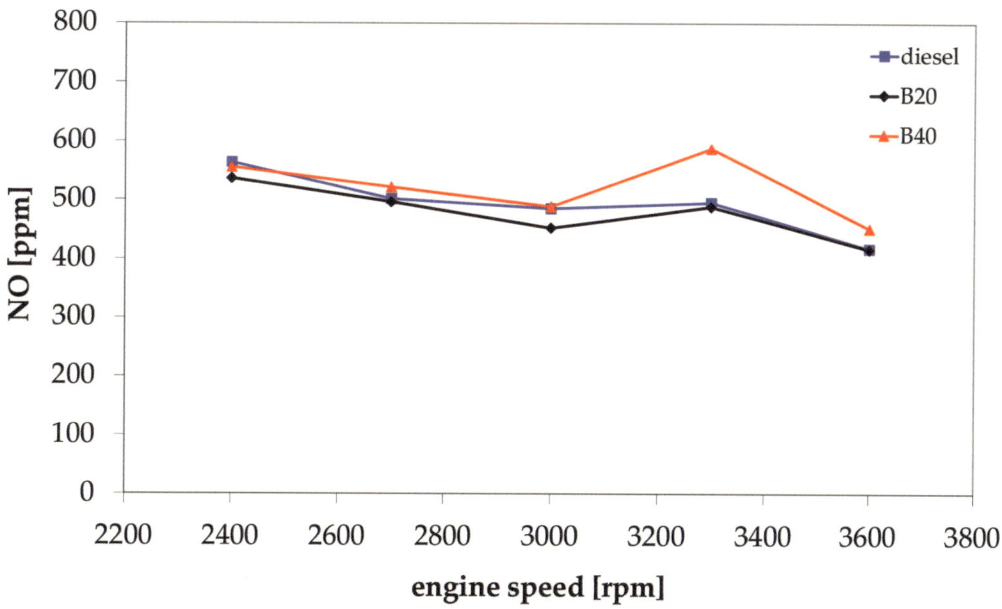

Figure 6. Variation of NO_x emission with engine speed [31].

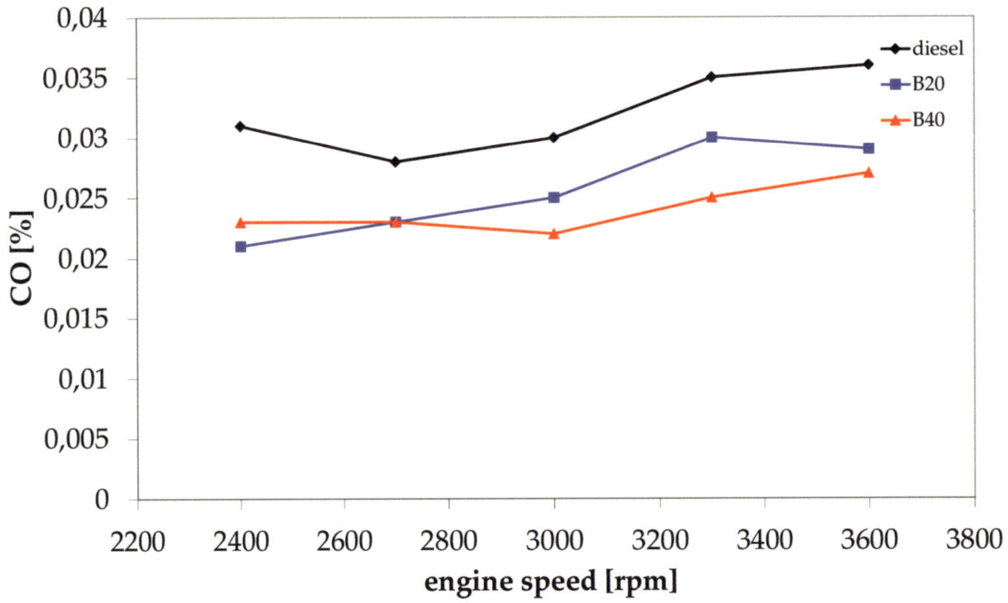

Figure 7. Variation of CO emission with engine speed [31].

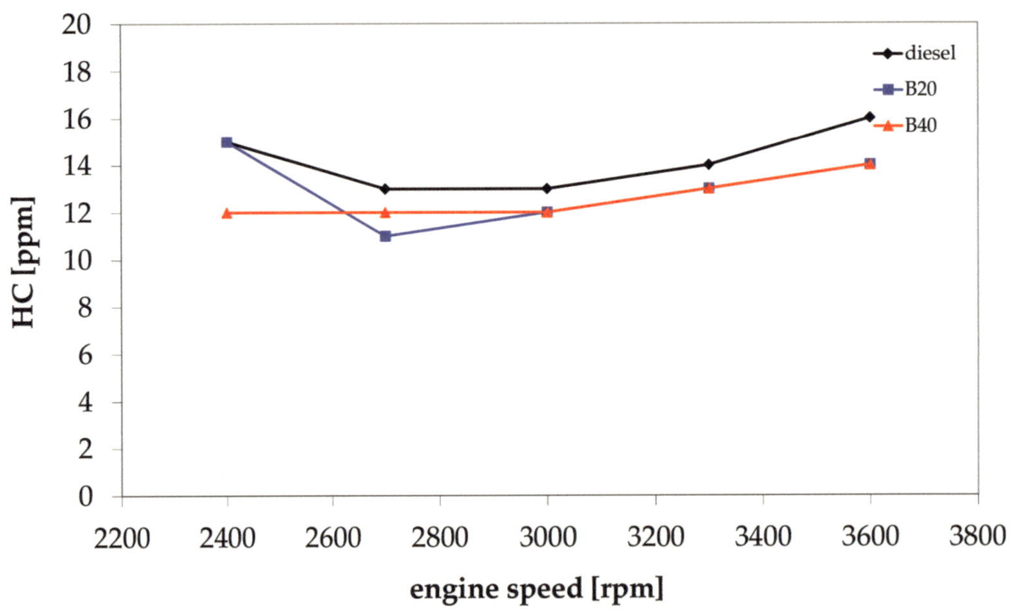

Figure 8. Variation of HC emission with engine speed [31].

an increase in CO_2 emission for biodiesel probably due to the higher density of biodiesel in comparison with diesel fuel, that increases the overall mass. Some studies [5, 23] report the opposite behavior as a consequence of the lower carbon to hydrogen ratio and the increase in oxygen content in the biodiesel blend.

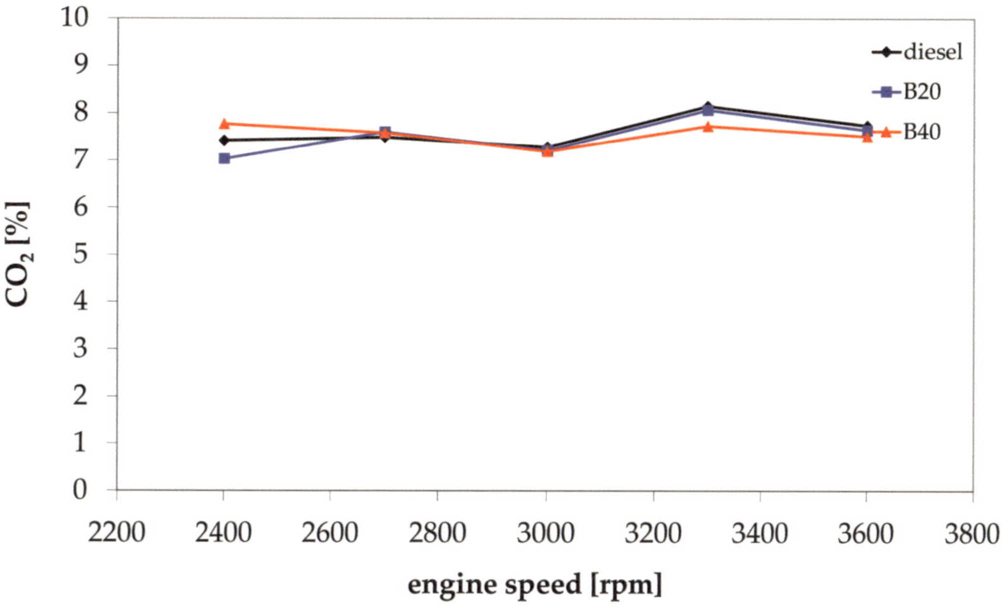

Figure 9. Variation of CO_2 emission with engine speed [31].

The following figures are devoted to analyze how the WCO content in the fuel affects the particle emission of the engine. **Figure 10** shows the nonvolatile particle number concentration (PNC) in the exhaust (all data have been normalized by the corresponding available engine output value). Each point is a cumulative value of particles in the range 23 nm–2.5 μm. The left-hand-side plot shows the variation of PNC with the engine speed at a fixed value of load (80%). The right-hand-side plot shows the effect of load condition at a fixed value of engine speed (3300 rpm). The traces highlight the reduction in soot emission obtained with B20 as regards diesel fuel. The increase in WCO ratio in the blend causes a further decrease in PNC. The obtained results are explained by accounting for many aspects. The higher density and viscosity of WCO blends as regards diesel oil are responsible for a variation of the injection process (smaller spray angle, larger droplet size, and fuel penetration length). Studies [12, 28] report that the injection setting has also a significant role in particle emission. Furthermore, the higher cetane number of WCO as regards diesel fuel causes a reduction in ignition delay and an increase in the mixing-controlled combustion duration. The higher oxygen content of biodiesel as regards ULSD promotes the combustion process and favors soot oxidation.

Figure 11 shows the effect of engine speed and load on soot concentration in the engine exhaust. The plots highlight the increase in the values as the engine speed increases for all tested fuels. Such a behavior is in agreement with similar results from literature [5, 19], and it is due to the concurrence aspects that have to be taken into account: a reduced time for air-mixing and combustion that penalizes the mixture uniformity and the combustion completeness. In addition, the increase in engine speed is responsible for an enhancement of the turbulence, which promotes the extent of complete combustion.

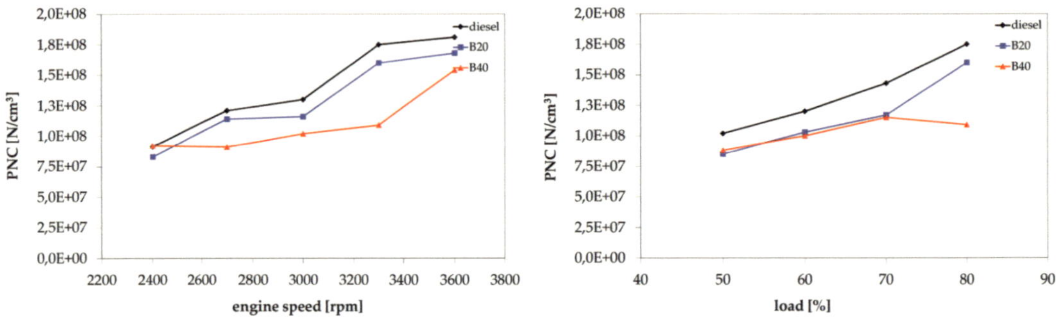

Figure 10. a): Particle number concentration at 80% load; b): particle number concentration at 3300 rpm [35].

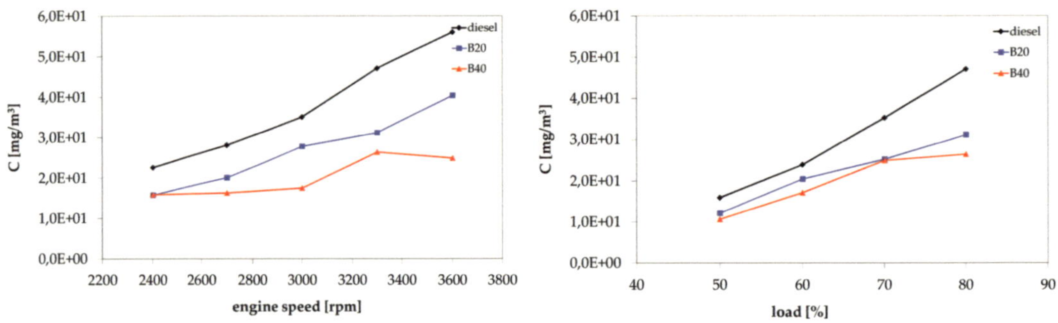

Figure 11. a): Soot concentration at 80% load, b): soot concentration at 3300 rpm [35].

The content of biodiesel in the fuel is responsible for a reduced particulate emission. This effect is ascribed to the increase in oxygen content in the blends that is responsible for a more complete combustion process and further oxidation of the already formed soot, according to Refs. [4–6, 8, 14, 19].

Figures 12–14 present the effect of blend ratio on the distribution of soot particles' diameters obtained during tests in which the engine was fuelled with ULSD, B20, and B40, respectively. In all plots, the data are expressed as size spectral density ($dN/dlogD_p/cc$). The left-hand-side plots show the variation of particle size obtained during tests in which the engine speed was varied at a fixed load condition (80%). The right-hand-side plots show the variation of particle size obtained during tests in which the load condition was varied at a fixed engine speed value (3300 rpm).

All trends exhibit a bimodal distribution of the particle size: 'nucleation' mode is comprised primarily of condensed volatile materials, mainly sulfate and heavy hydrocarbons, with particle sizes that are typically less than 30 nm; 'accumulation' mode is comprised mainly of carbonaceous particles of sizes larger than 30 nm [36]. The engine type, the operation condition, and the dilution needed prior sampling deeply affect the particle size distribution [4, 24].

The graphs highlight that accumulation mode dominates in all tested conditions. ULSD shows a decrease in particle diameters as engine speed increases: the number of particles of

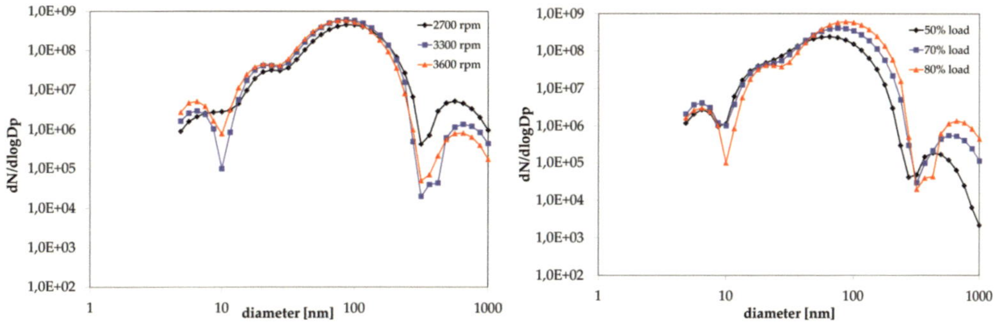

Figure 12. a): Particle number concentration at 80% load for ULSD, b): particle number concentration at 3300 rpm for ULSD [35].

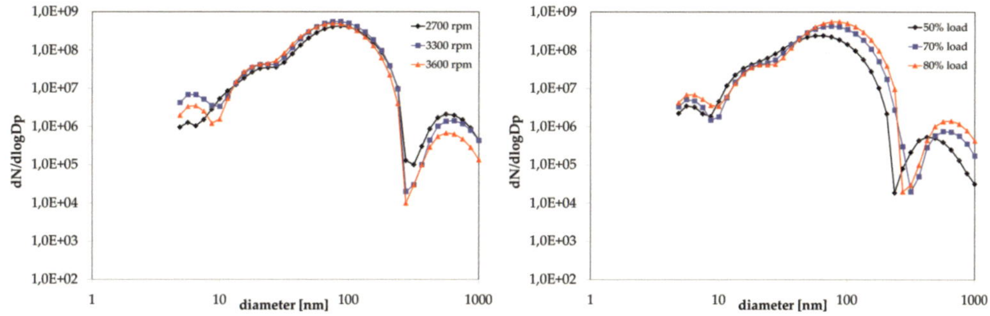

Figure 13. a): Particle number concentration at 80% load for B20, b): particle number concentration at 3300 rpm for B20 [35].

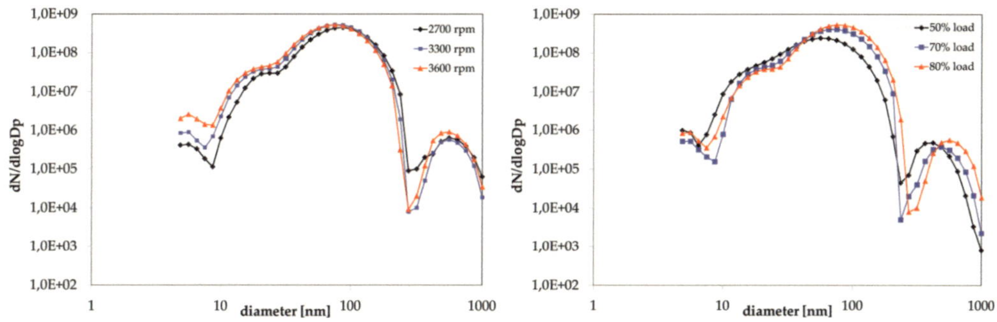

Figure 14. a): Particle number concentration at 80% load for B40, b): particle number concentration at 3300 rpm for B40 [35].

diameter larger than 100 nm decreases; the number of particles of diameter lower than 100 nm increases. Load condition affects the particle size distribution; load increase causes a greater number of larger particles, in agreement with Ref. [14].

Figures 13 and **14** show the data obtained with B20 and B40, respectively. The engine operative conditions affect the particle size distribution. B20 traces agree with those related to diesel fuel; the increase in engine speed causes a decrease in soot particle concentration with diameters

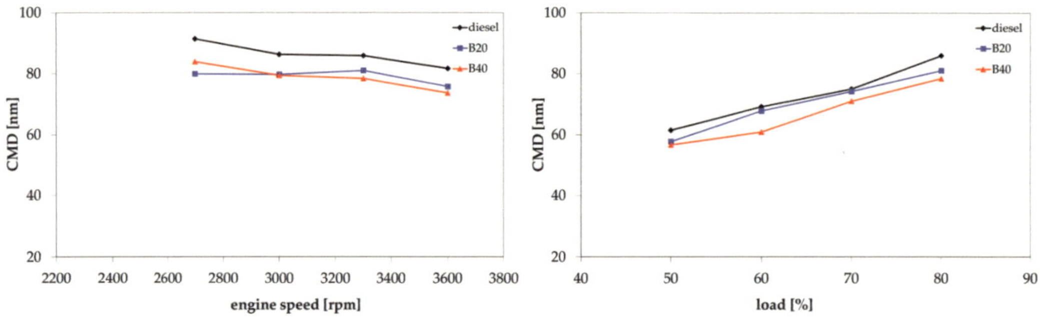

Figure 15. a): Accumulation mean diameter at 80% load, b): accumulation mean diameter at 3300 rpm [35].

larger than 100 nm. B40 trends show that the increase in engine speed is responsible for an increase in the concentration of particles of diameters under 100 nm and larger than 200 nm. The traces highlight the abrupt decrease in particle number concentration in the range of diameters around 1 μm. The increase in load (at constant engine speed) is responsible for the increase in particles' sizes for both B20 and B40.

The comparison between the particles' distributions points out that B20 and B40 are characterized by lower number of particles than ULSD in almost all diameters. This behavior agrees with literature data. Studies report that the employment of biodiesel blends produces an increased number of nanoparticles and a reduced number of ultrafine and fine particles in comparison with ULSD [6, 37, 38]. It can be explained by the oxygen content of WCO that favors the combustion completeness in the region of fuel-rich diffusion flame and then promotes the oxidation of the already-formed soot and inhibits the soot growth [19].

In **Figure 15**, the variation with engine operative conditions of the mean size of accumulation mode is shown. For all tested fuels, the court mean diameter decreases with increasing engine speed at constant load value. The increase in load condition causes an increase in the court mean diameter.

WCO biodiesel blends have lower mean diameter in their exhaust than diesel fuel. B40 has smaller particle mean diameter as regards B20 for almost all tested conditions.

4. Conclusion

Biodiesel from waste cooking oil was tested to investigate the impact of WCO percentage in blends with ultralow-sulfur diesel oil in the performance and emission characteristics of an up-to-date light and compact common rail diesel engine, whose main application is in microcars and in urban vehicles.

The engine performance (in terms of available torque, brake specific fuel consumption, brake thermal efficiency) with biodiesel blends was found quite similar for B20 and diesel fuel. B40 suffered for the lower caloric value of the blend in comparison with ULSD.

For what concerns the exhaust emissions, a reduction in CO and HC was observed for biodiesel blends, which was more significant with the increase in WCO in the fuel.

A reduction in particulate emissions was attained for WCO blends, along with a corresponding increase in NO_x, according to the well-known trade-off between NO_x and particulate. Particle size distributions were characterized by a bimodal distribution, in which accumulation mode dominated. A slight reduction in the particulate number concentration was observed as compared to diesel oil. The reduction was more evident as the WCO content in the fuel increased. The mean size of particles in B20 and B40 was smaller than that obtained with ULSD. For all fuel, engine load and speed conditions affected the particle size distribution: the increase in engine speed was responsible for a reduction in particles' diameters; the increase in load led to a reduction in the number of smaller particles.

Acknowledgements

We acknowledge the fundamental contribution of AVL which provided the instrumentation for the particle matter measurements (AVL Particle Counter and AVL Micro Soot Sensor) used during the research activity.

Nomenclature

B20	80% ULSD and 20% WCO, by volume
B40	60% ULSD and 40% WCO, by volume
BSFC	brake specific fuel consumption
BTE	brake thermal efficiency
CMD	count mean diameter
D_p	particle diameter
FFA	free fatty acids
PNC	nonvolatile particle number concentration
ULSD	ultralow-sulfur diesel
WCO	waste cooking oil

Author details

Giancarlo Chiatti*, Ornella Chiavola and Erasmo Recco

*Address all correspondence to: ornella.chiavola@uniroma3.it

Engineering Department, 'ROMA TRE' University, Rome, Italy

References

[1] Demirbas A. Progress and recent trends in biodiesel fuels. Energy Conversion and Management. 2009;**50**:14-34. DOI: 10.1016/j.enconman.2008.09.001

[2] Peng D. Exhaust emission characteristics of various types of biofuels. Advances in Mechanical Engineering. 2015;**7**(7):1-7. DOI: 0.1177/1687814015593036

[3] Puzun A, Wanchen S, Guoliang L, Manzhi T, Chunjie L, Shibao C. Characteristics of particle size distributions about emissions in a common-rail diesel engine with biodiesel blends. Procedia Environmental Sciences. 2011;**11**(C):1371-1378. DOI: 10.1016/j.proenv.2011.12.206

[4] Lapuerta M, Armas O, Rodriguez-Fernandez J. Effect of biodiesel fuels on diesel engine emissions. Progress in Energy Combustion Science. 2008;**34**(2):198-223. DOI: 10.1016/j.pecs.2007.07.001

[5] Xue J, Grift TE, Hansen AC. Effect of biodiesel on engine performances and emissions. Renewable and Sustainable Energy. 2011;**15**:1098-1116. DOI: 10.1016/j.rser.2010.11.016

[6] Kim H, Choi B. The effect of biodiesel and bioethanol blended diesel fuel on nanoparticles and exhaust emissions from CRDI diesel engine. Renewable Energy. 2010;**35**:157-163. DOI: 10.1016/j.renene.2009.04.008

[7] Kegl B. Influence of biodiesel on engine combustion and emission characteristics. Applied Energy. 2011;**88**(5):1803-1812. DOI: 10.1016/j.apenergy.2010.12.007

[8] Buyukkaya E. Effects of biodiesel on a DI diesel engine performance, emission and combustion characteristics. Fuel. 2010;**89**:3099-3105. DOI: 10.1016/j.fuel.2010.05.034

[9] Song H, Tompkins BT, Bittle JA, Jacobs TJ. Comparisons of NO emissions and soot concentrations from biodiesel-fuelled diesel engine. Fuel. 2012;**96**:446-453. DOI: 10.1016/j.fuel.2012.01.004

[10] Devendra S, Subramanian KA, Singal KS. Emissions and fuel consumption characteristics of a heavy duty diesel engine fueled with hydroprocessed renewable diesel and biodiesel. Applied Energy. 2015;**155**:440-446. DOI: 10.1016/j.apenergy.2015.06.020

[11] Serrano L, Loopes M, Pires N, Ribeiro I, Cascão P, Tarelho L, Monteiro A, Nielsen O, Gameiro da Silva M, Borrego C. Evaluation on effects of using low biodiesel blends in a EURO 5 passenger vehicle equipped with a common-rail diesel engine. Applied Energy. 2015;**146**:230-238. DOI: 10.1016/j.apenergy.2015.01.063

[12] Attia AMA, Hassaneen AE. Influence of diesel fuel blended with biodiesel produced from waste cooking oil on diesel engine performance. Fuel. 2016;**167**:316-328. DOI: 10.1016/j.fuel.2015.11.064

[13] Sun J, Caton JA, Jacobs TJ. Oxides of nitrogen emissions from biodiesel-fuelled diesel engines. Progress in Energy and Combustion Science. 2010;**36**(6):677-695. DOI: 10.1016/j.pecs.2010.02.004

[14] Agarwal AK, Gupta T, Kothari A. Particulate emissions from biodiesel vs diesel fuelled compression ignition engine. Renewable and Sustainable Energy Reviews. 2011;**15**(6):3278-3300. DOI: 10.1016/j.rser.2011.04.002

[15] Hwang J, Jung Y, Bae C. Spray and combustion of waste cooking oil biodiesel in a compression-ignition engine. International Journal of Engine Research. 2015;**16**(5):664-679. DOI: 0.1177/1468087415585282

[16] Szybist JP, Song J, Alam M, Boehman AL. Biodiesel combustion, emissions and emission control. Fuel Processing Technology. 2007;**88**:679-691. DOI: 10.1016/j.fuproc.2006.12.008

[17] Rakopoulos CD, Antonopoulos KA, Rakopoulos DC, Hountalas DT, Giakoumis EG. Comparative performance and emissions study of a direct injection diesel engine using blends of diesel fuel with vegetable oils or bio-diesels of various origins. Energy Conversion & Management. 2006;**47**:3272-3287. DOI: 10.1016/j.enconman.2006.01.006

[18] Lee CS, Park SW, Kwon S. An experimental study on the atomization and combustion characteristics of biodiesel-blended fuels. Energy & Fuels. 2005;**19**:2201-2208. DOI: 10.1021/ef050026h

[19] Yehliu K, Boehman AL, Armas O. Emissions from different alternative diesel fuels operating with single and split fuel injection. Fuel. 2010;**89**(2):423-437. DOI: 10.1016/j.fuel.2009.08.025

[20] Ajtai T, Pintér M, Utry N, Kiss-Albert G, Gulyás G, Pusztai P,Puskás R, Bereczky A, Szabados G, Szabó G, Kónya Z, Bozóki Z. Characterisation of diesel particulate emission from engines using commercial diesel and biofuels. Atmospheric Environment. 2016;**134**:109-120. DOI: 10.1016/j.atmosenv.2016.03.046

[21] Phan AN, Phan TM. Biodiesel production from waste cooking oil. Fuel. 2008;**87**(17-18):3490-3496. DOI: 10.1016/j.fuel.2008.07.008

[22] Gopal KN, Arindan P, Sharma S, Samanchi C, Sathyanarayanan K, Elango T. Investigation of emission and combustion characteristics of a IC engine fuels with waste cooking oil methyl ester and diesel blends. Alexandria Engineering Journal. 2014;**53**(2):281-287. DOI: 10.1016/j.aej.2014.02.003

[23] An H, Yang WM, Chou SK, Chua KJ. Combustion and emission characteristics of diesel engine fueled by biodiesel at partial load conditions. Applied Energy. 2012;**99**:363-371. DOI: 10.1016/j.apenergy.2012.05.049

[24] An H, Yang WM, Maghbouli A, Li J, Chou SK, Chua KJ. Performance, combustion and emission characteristics of biodiesel derived from waste cooking oils. Applied Energy. 2013;**112**:493-499. DOI: 10.1016/j.apenergy.2012.12.044

[25] Can O. Combustion characteristics, performance and exhaust emissions of a diesel fueled with a waste cooking oil and biodiesel mixture. Energy Conversion and Management. 2014;**87**:676-686. DOI: 10.1016/j.enconman.2014.07.066

[26] Cheung CS, Man XJ, Fong KW, Tsang OK. Effect of waste cooking oil biodiesel on the emissions of a diesel engine. Energy Procedia. 2015;**66**:93-96. DOI: 10.1016/j.egypro.2015.02.050

[27] Man XJ, Cheung CS, Ning Z. Effect of diesel engine operating conditions on the particulate size, nanostructure and oxidation properties when using wasting cooking oil biodiesel. Energy Procedia. 2015;**66**:37-40. DOI: 10.1016/j.egypro.2015.02.020

[28] Hwang J, Bae C, Gupta T. Application of waste cooking oil (WCO) biodiesel in a compression ignition engine. Fuel. 2016;**176**:20-31. DOI: 10.1016/j.fuel.2016.02.058

[29] Chiatti G, Chiavola O, Recco E, Palmieri F. Soot particles experimental characterization during cold start of a micro car engine. Energy Procedia. 2016;**101**:662-669. DOI: 10.1016/j.egypro.2016.11.084

[30] Chiatti G, Chiavola O, Palmieri F, Piolo A. Diagnostic methodology for internal combustion diesel engines via noise radiation. Energy Conversion & Management. 2015;**89**:34-42. DOI: 10.1016/j.enconman.2014.09.055

[31] Chiatti G, Chiavola O, Palmieri F, Albertini S. Combustion and emissions characterization of biodiesel blends in a city-car engine. Energy & Fuels. 2014;**28**(8):5076-5085. DOI: 10.1021/ef501023q

[32] Altun S, Bulut H, Oner C. The comparison of engine performance and exhaust emission characteristics of sesame oil-diesel fuel mixture with diesel fuel in a direct injection diesel engine. Renewable Energy. 2008;**33**:1791-1795. DOI: 10.1016/j.renene.2007.11.008

[33] Qi DH, Chen H, Geng LM, Bian YZH, Ren XCH. Performance and combustion of biodiesel-diesel-methanol blend fuelled engine. Applied Energy. 2010;**87**:1679-1686. DOI: 10.1016/j.apenergy.2009.10.016

[34] Ng JH, Ng HK, Gan SY. Engine-out characterization using speed-load mapping and reduced test cycle for a light-duty diesel engine fuelled with biodiesel blends. Fuel. 2011;**90**:2700-2709. DOI: 10.1016/j.fuel.2011.03.034

[35] Chiatti G, Chiavola O, Recco E, Palmieri F. Impact of waste cooking oil in biodiesel blends on particle size distributions from a city-car engine. Journal of Energy Institute. DOI: 10.1016/j.joei.2016.11.009

[36] Myung CL, Park S. Exhaust nanoparticle emissions from internal combustion engines: A review. International Journal of Automotive Technology. 2012;**13**(1):9-22. DOI: 10.1007/s12239-012-0002-y

[37] Neer A, Koylu UO. Effect of the operating conditions on the size, morphology, and concentration of sub-micrometer particulates emitted from a diesel engine. Combustion and Flame. 2006;**146**:142-154. DOI: 10.1016/j.combustflame.2006.04.003

[38] Tsolakis A. Effects on particle size distribution from the diesel engine operating on RME-biodiesel EGR. Energy & Fuels. 2006;**20**:1418-1424. DOI: 10.1021/ef050385c

Chapter 3

n-Butanol-Diesel (D2) Blend Fired in a Turbo-Charged Compression Ignition Engine: Performance and Combustion Characteristics

Lennox Siwale, Lukacs Kristof, Torok Adam,
Akos Bereczky, Makame Mbarawa,
Antal Penninger and Andrei Kolesnikov

Additional information is available at the end of the chapter

http://dx.doi.org/10.5772/intechopen.72879

Abstract

The use of biofuels that include n-butanol in diesel fuel (DF) is attracting attention in the search for the reduction of emissions into the environment due to the burning of fossil fuel. The performance and combustion characteristics were evaluated in this study using blends B5, B10, and B20 (B5: 5% n-butanol and 95% DF) in a turbo-charged direct injection compression ignition engine. In the n-butanol diesel studies, a comparison was made with other studies that also included biodiesel in order to determine how suitable n-butanol-diesel blends were to use in internal combustion engines. Combustion characteristics of B20 (n-butanol 20% and 80% DF) improved when the study was compared with a similar study that included 40% biodiesel added to B20. A higher value of the standard deviation for DF than the blends was observed from the standard deviation diagram, indicating a more stable combustion process for the blends than DF. Soot reduction relative to DF at 1500 rpm at 75% load for B05, B10, and B20 mixtures was 55.5, 77.8, and 85.1%, respectively. This reduction is a significant advantage of blending DF with smaller shared volumes of bioalcohol.

Keywords: compression ignition, n-butanol-diesel, performance, combustion

1. Introduction

A very strong debate on the gradual substitution of petroleum by using renewable alternatives such as biofuel dominates the political and economic agenda worldwide [1]. This has been driven by the global concern of the international governing bodies for countries to comply with the requirements for emission reduction. The energy independence and security act of 2007 (US) requires the production of 36 billion gallons of renewable fuels including ethanol-based fuels by 2022. This constitutes a 75% increase in their production over the next 14 years [2].

Biofuels constitute a renewable source of energy derived from biomass (the primary source). These include all three states of matter: solid, liquid and gas. They fall into the following categories, for liquid biofuels: (a) bioalcohols, (b) vegetable oils and biodiesel, and (c) biocrude and synthetic oils. It is expected that their demand in the future will rise because they replace petroleum [3]. Biofuels can be produced from bio-based materials through the biochemical processes such as pyrolysis, gasification, or liquefaction. Thermal chemical reforming of biomass concerns the process of catalytic or non-catalytic pyrolysis and gasification with a goal of maximizing in the production of energetically exploitable liquids or gases. Biofuels include bioethanol and biomethanol, Fischer-Tropsch liquids and biogas [4].

Oxygenated fuels including alcohols constitute one of the categories of the biofuels that have attracted research for many years due to their cleaner properties when blended with the conventional fuels [5]. Oxygenated fuels drew the attention of many researchers, due to their ability to drastically reduce the particulate matter (PM) emission without altering the emission levels of NO_x, UHC, and CO [6, 7]. Oxygenated additives such as ethers, esters, alcohols have been added to diesel fuel [8]. Alcohols are used in internal-combustion engines as alternative fuels. If the source of alcohol is based on biomass then one can reduce greenhouse gas emissions, and exhaust emissions as well as enhance the overall energy efficiency [1, 3, 9–11].

N-butanol possesses certain qualities that make it more suitable to use in internal combustion engines than ethanol or methanol. These include a higher heating value of (28.4 MJ/L) for n-butanol and (21.2 MJ/L) for ethanol; higher cetane number (CN), lower auto-ignition temperature, better flow ability. For ethanol-diesel blend, an emulsifying agent is used to satisfy the homogeneity of the ethanol-diesel fuel mixture to prevent phase separation [12]. However, n-butanol is easily miscible in diesel fuel [13]. Within the functional groups of butanol isomers, n-butanol has the highest flame speed which enhances the combustion process [14]. Some researchers have argued that n-butanol-diesel blends prolong the ignition delay (ID), thereby affecting the combustion duration in internal combustion engines. However, others have stated that burning n-butanol-diesel fuel blends reduces emissions [14]. Therefore, comparisons are necessary on how the combustion characteristics differ between blends of n-butanol and biodiesel in diesel fuel, respectively.

In internal combustion engines such as diesel engines, diesel fuel is atomized into droplets using injectors at high pressure. These droplets entrain air as they evaporate into proportions that are combustible in the combustion chamber. The air temperature and pressure are above the fuel's ignition point. Therefore, after a short while: or ID the mixture auto ignites. ID is a time lag that is influenced by the physical chemical properties of the fuels measured in time (s) or CADs in engine cycles. This affects the combustion process in internal combustion engines including combustion duration and in-cylinder pressure and temperatures.

The simultaneous use of biodiesel, ethanol, and butanol in diesel is an interest that has arisen recently by researchers. However, since biodiesel is characterized by high viscosity, lubricity,

CN, and flash point relative to that of ethanol, the obstacles due to ethanol are partially overcome. Literature concerning the use of n-butanol/diesel fuel blends in diesel engines and their effect on the steady state performance and exhaust emission in engines is limited [15]. Most of the studies, however, have evaluated the potential of methanol and ethanol [16], while there is a need for more studies to be done on the characteristics of higher alcohols such as propanol, butanol, and pentanol [17].

2. Objectives

The general objective is to reduce the negative impacts of petroleum oil-based fuels in reciprocating engines on the environment through the use of oxygenated (alcohol) blends, without deteriorating the engine performance. The specific objective of the study is as follows:

- To evaluate the performance and combustion characteristics of n-butanol-diesel blends: B5, B10, and B20. It is given in the abstract in a direct-injection, turbo-charged diesel engine.
- To compare findings with studies that were carried out by others [18–20].

3. Engine experimentation

In this section the engine experimental set up, procedure and matrix are included.

The study was conducted on a four-cylinder piston 1Z 1.9 L-66 kW Turbo-Direct injection (TDI) Wolkswagen diesel engine. The engine was fully equipped with a turbocharger with all the necessary sensors and actuators for stable operation of engine. The engine was warmed up to about 94°C and the temperature was maintained. The engine was run for 20–30 min to warm up with the reference DF or test blend fuel. Once the new fuel was pumped in, the engine was made to run for about 20 min to allow for stable operation of the new blend at test fuel conditions. Typical fuel consumption data and brake specific consumption are shown in **Table 1**. The fuel injection timings for the blends and DF controlled by the electronic diesel control (EDC) for different speeds of 1500, 2500, 3000, and 3500 rpm were 11,11,12, and 15° crank angle BTDC, respectively. The fuel consumption was measured using the AVL 7030 dynamic fuel consumption measuring equipment. Torque was measured by a Borghi and Saveri FE-350S eddy current dynamometer. The study was carried out for 75, 50, and 25% load for engine operating a range of 1500, 2500, 3000, and 3500 rpm. The engine was made to run for each measuring point on steady-state condition for about 2 min before recording values. An average based on three measurements for each test fuel was recorded. The engine operating conditions are shown in **Table 2**. The in-cylinder indicated pressure (IP) was measured by a piezoelectric pressure transducer Kistler KIAG 600. The top dead center (TDC) crank angle position was measured by an Optical Encoder HENGSTLER RI 32-0/1024.ER.14KA [21].

	DF	B05	B10	B20
Fuel cons. (kg/h)				
(A) At 1500 rpm	5.47	5.47	5.43	5.63
(B) At 1500 rpm	3.0	3.06	3.12	3.12
(B) At 3000 rpm	7.13	7.34	7.56	7.62
λ (−)				
(A) At 1500 rpm	1.4	1.42	1.5	1.52
(B) At 1500 rpm	1.72	1.82	1.92	1.92
(B) At 3000 rpm	2.37	2.73	2.42	2.79
BSFC (g/kWh)				
(A) At 1500 rpm	235	231	237	248
(B) At 1500 rpm	252.9	258.02	263.09	263.08
(B) At 3000 rpm	249.4	256.75	264.4	266.54

Table 1. Typical fuel consumption, λ, BSFC with DF and blends at 1500 and 3000 rpm; A = 100%, and B = 75% load [21].

Torque (Nm)	Speed (rpm)	Exhaust temperature (°C)	*MAP (bar)
184.6	3004	562	0.82
136.4	3004	473	0.71
92	3002	419	0.56
45	3000	333	0.38
168.9	3504	584	0.79
187.8	2505	519	0.84
153.7	1502	532	0.35

Ambient air temperature 20°C, intake air temperature 25°C.
*Manifold (boost) air pressure (above atmospheric pressure).

Table 2. Operating conditions of engine for fuel D2 [21].

Typical fuel consumption and fuel air ratios are given in **Table 1** and operating conditions in **Table 2**.

4. Diesel (TDI) engine

4.1. Crank angle advance injection timing

The effect of injection timing of all the test fuels (B0, B5, B10, and B20) is indicated in **Figure 1**. The engine operating speeds selected for testing were 1500 rpm and 3000 rpm. When the operating speed was high the fuel was injected earlier than when the speed was reduced to allow for the correct retention time for the fuel to burn adequately. Running on partial load in

http://dx.doi.org/10.5772/intechopen.72879

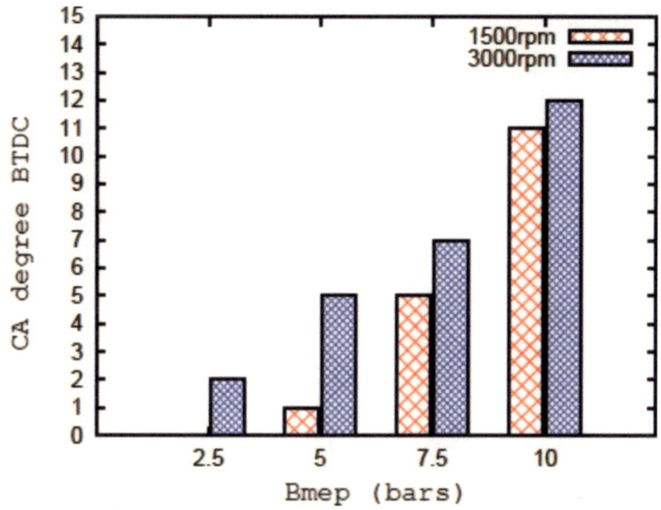

Figure 1. Effect of test fuels on injection timing.

terms of brake-mean effective pressure (BMEP), or when the engine was running at low speed, the engine's start of injection (SOI) timing was retarded by the EDC in order to meet the engine's operating conditions.

4.2. Manifold boost (air) pressure

Figure 2a and **b** depict the effect of the blends on manifold boost (air) pressure at 1500 rpm and 3000 rpm, respectively. The increasing boost (air) pressure (above atmospheric of 1 bar) level with BMEP is a measure that helps to improve the brake thermal efficiency (BTE) as the fuel-air ratio is reduced.

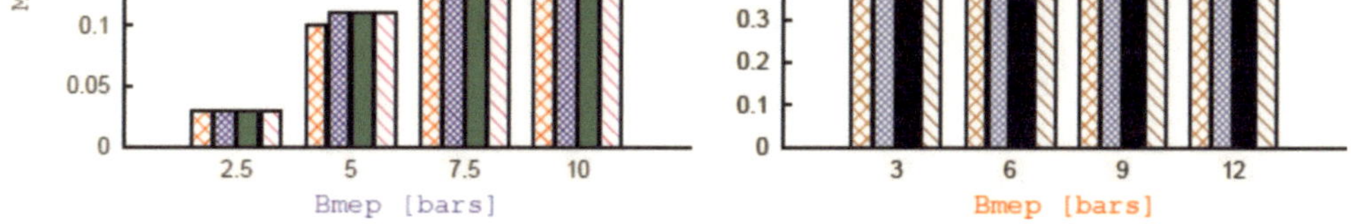

Figure 2. Manifold boost (air) pressure (MAP) vs. BMEP (a) at 1500 rpm (b) at 3000 rpm.

4.3. Torque, power, and EGT

The derated power and torque indicated in **Figure 3** is attributed to the fixed algorithm and set points in the EDC unit, which reads diesel as the reference fuel as programmed if not modified. In this case, the EDC was not calibrated for the n-butanol/diesel blends. When the diesel fuel was changed to n-butanol/diesel blend, the lower heating value due to the blend triggered the EDC to increase the mass flow of fuel to compensate for the drop in energy delivered; the heating value of n-butanol is lower than that of diesel fuel. This is achieved by altering the fuel-control ring position of the control system. However, the EDC set point for DF which controls the amount of diesel that can be delivered at different loads was fixed. Therefore, the mass flow of the blend could not be increased above this point. This explains the lowered power and torque output when using n-butanol/DF as shown in **Figure 3**.

The effect of exhaust gas temperature (EGT) on BMEP is indicated in **Figure 4a** and **b**. The trend shows an increase of EGT with BMEP for all the test fuels: DF, B5, B10, and B20. The

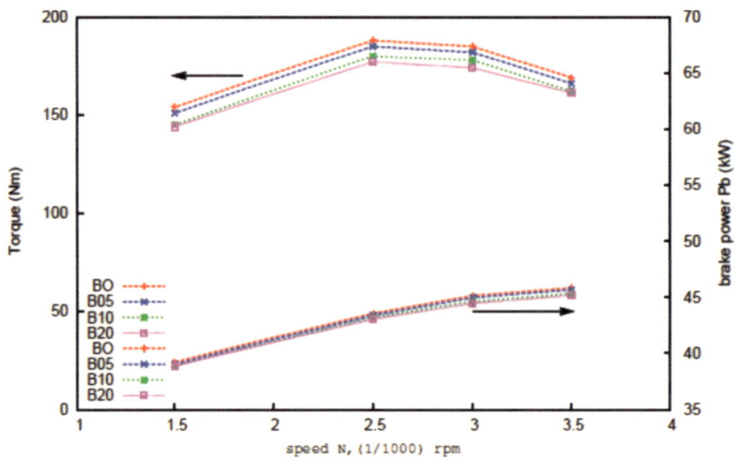

Figure 3. Effects of blends on torque, brake power vs. speed [21].

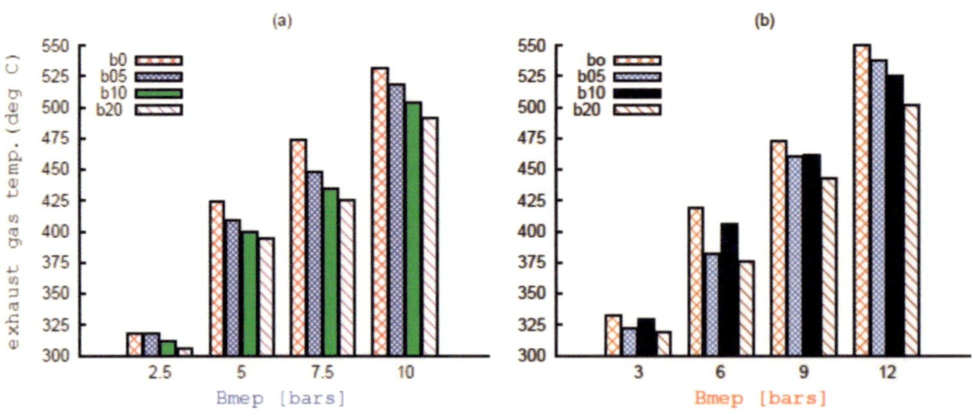

Figure 4. EGT vs. BMEP (a) at 1500 rpm and (b) at 3000 rpm [21].

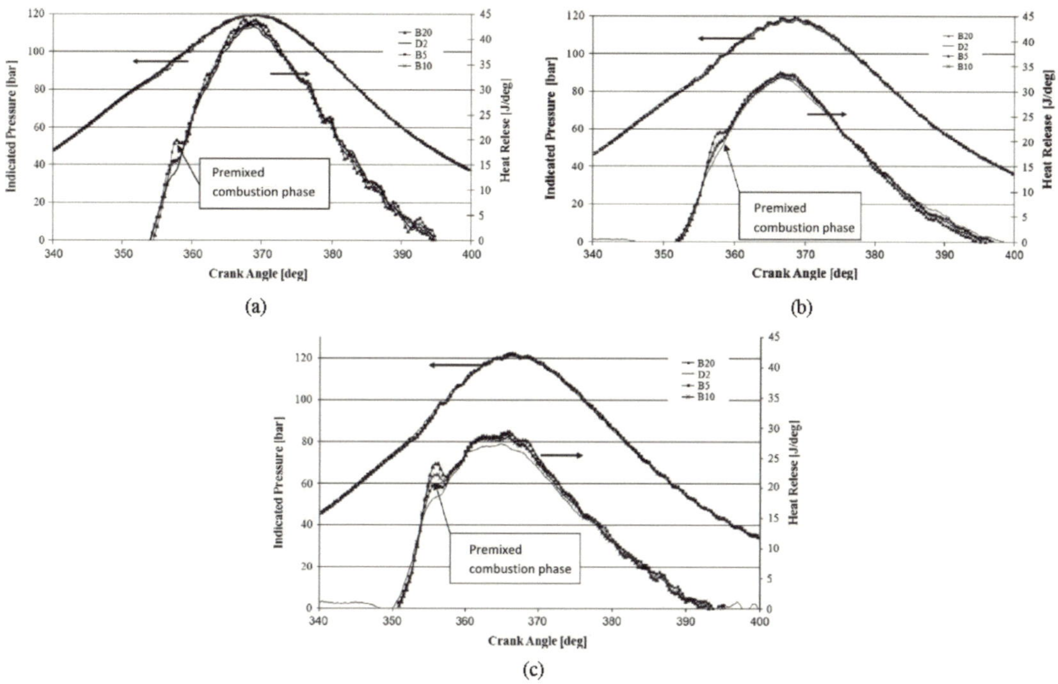

Figure 5. (a) Indicated pressure (ip) and HRR at 2500 rpm source [21]; (b) indicated pressure and HRR at 3000 rpm source [21]; (c) indicated pressure and HRR at 3500 rpm source [21].

higher heat of evaporation of the blends than DF resulted in evaporative cooling which lowers EGT. Furthermore, the increasing higher molecular oxygen content of the blends than DF is also a contributing factor in lowering EGT by reducing the energy content of the fuel. The irregularity for B10 at 25% and 50% load at 3000 rpm could be due to unstable combustion, which is also evident in the high standard of deviation for B10% in **Figure 6**.

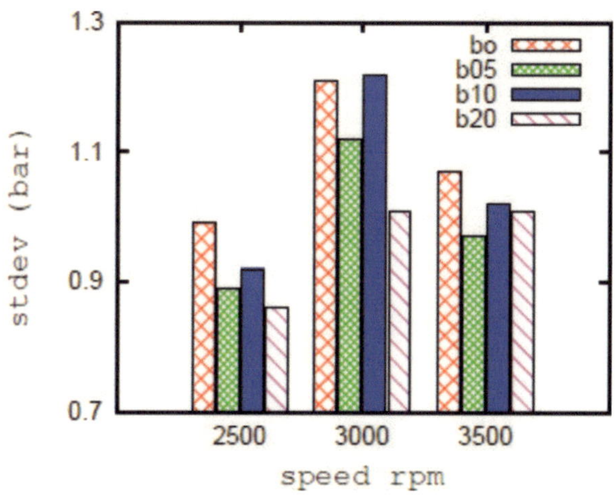

Figure 6. Standard deviation of indicated pressure [21].

5. Indicated pressure, HRR, and effect of blends on (TDI) CI engine

In combustion studies, the CN measures the ease with which a fuel auto-ignites. The CN is therefore a measure of a fuel's auto-ignition delay, the period between the start of injection and the first identifiable pressure increase during the combustion of the fuel. The heat release is the net heat of combustion generated by chemical reactions of the fuel affected by the fuel's physical and chemical properties in the combustion process of the fuel. The fuel is injected timely and periodically in the combustion chamber by means of fuel injectors. The mean pressure developed due to the heat released each cycle is termed as the IP, which is the pressure reading on an indicator diagram.

Figure 5a–c illustrates the effect of the mean pressure and the heat release rate at different speeds of 2500, 3000, 3500 rpm on blends compared with DF. The operating conditions depicted by brake mean effective pressure (BMEF) for the blends is given in **Table 3**. It is observed that there are very little differences in the peak pressures obtained at the different speeds. After the Ignition delay, there exists a rapid combustion phase that causes a steep rise in the in-cylinder pressure and heat release. This is known as the premixed phase where after atomization of the fuel, the fuel droplets are evaporated and ignited by the heated air from the compression process prior to the start of injection. The peak pressure is controlled by the ignition delay, the rate of fuel injection, the speed of the engine and including the compression ratio for the engine. The remaining oxygen of the air charge limits the final phase slowing down the combustion rate (see **Figure 5a–c** heat release curve) as the crank angles advance toward the beginning of exhaust process. Increasing the shared volume of n-butanol in DF as well as the engine speed coupled with the molecular oxygen content of the oxygenated fuel improves the mixing quality of the blend. This phenomenon is observed in the increasingly distinguishable premixed combustion phase with an increase of engine speed (see **Figure 5b** and **c** heat release curve). The premixed phase is similar for B5 and B10 at 2500 and 3000 rpm. Higher heat release rate is observed at 2500 rpm because of the more resident time (s) allowed for the mass of fuel to burn than at higher speeds of 3000 and 3500 rpm. The mixing controlled combustion or final phase just after the premixed stage indicated a steeper slope on the heat release curve for the blends than DF with increasing the shared volume of n-butanol in the blend. By this, the combustion efficiency of the blends was better than DF and the combustion duration shortened.

In past studies conducted by Ref. [19], the effect of using BD30 (30% biodiesel or methyl ester, from rapeseed oil (RME) blended with 70% diesel fuel) alone shortened the ignition delay (ID). By adding E10 (10% ethanol and 90% diesel fuel) in other words, BD30 + E10 prolonged the ID,

Speed [rpm]	B0	B05	B10	B20
2500	12.45	12.25	11.92	11.76
3000	12.23	12.09	11.8	11.50
3500	11.19	10.98	10.74	10.64

Table 3. Operating BMEP (bars) for **Figure 5a–c** [21].

which is not desirable as it results in a hard start of the test engine. When comparing the CNs, it was found that the CN for RMEs was 55 whereas for DF was 51. The CN for n-butanol was approximately 25 and CN for ethanol was approximately 8. Therefore, increasing the shared volume of n-butanol above 20% in Diesel fuel displaces richer components of CN in DF than in n-butanol prolonging start of combustion SOC and prompting the hard start of the engine. Therefore, this limits how many fractions of n-butanol can be blended with DF. A blend of greater than 20% shared volume of n-butanol in DF further reduces the CN of the blend. Therefore, the limiting point would be when the final CN of the blend is too low for the engine to start.

The combustion characteristics of blend B20 in the author's study is compared with Bu20 (n-butanol 20%, biodiesel 40%, and diesel 40%) by others [20]. In the latter study, B50 (biodiesel 50% and diesel fuel 50%) was the preferred fuel type over Bu20 in terms of maximum pressure and heat release rate. The operating conditions were: single cylinder, four strokes, direct injection, air cooled, TecQuipment TD212 diesel engine, naturally aspirated, maximum power 3.5 kW at 3600 rpm and maximum heat release rate (HRR) 15.5 J/deg. In the author's study using B20, the maximum HRR was 43.5 J/deg. for four cylinders, four strokes, direct injection, turbo-charged diesel engine as specified in section (1). In a study by others, B50 was recommended as the most suitable blend fired in a diesel engine. Combustion duration for DF is higher (within the range for the indicated power of 1.2 and 1.6 kW) than Bu20. Combustion duration in the author's study using B20 (n-butanol 20%, DF 80%) was reduced as observed by a steeper profile of the heat curve in comparison with DF (D2) see **Figure 5a–c**. **Figure 5a–c** illustrate the mean IP which did not change much in terms of peak pressure for the increase of HRR at lower speeds for all test fuels. The IP and HRR were evaluated at 2500, 3000, and 3500 rpm for the entire test fuels at full load respectively. The operating BMEP for the blends are presented in **Table 3**. It was observed that at 3500, 3000 and 2500 rpm within 5° CAD the premixed heat release was within 17–24 J/deg., 15–23 J/deg., and 10–20 J/deg., respectively. In the study by others [20], the maximum pressure was obtained by DF followed by B50 and then Bu20 [20] at a constant speed of 1500 rpm. This could be attributed to the effect of the EDC which could restrict the fuel flow of the blends when not modified (see Section 4.3).

These results concur with the findings of past study conducted by [19], who obtained a prolonged ID by adding an alcoholic admix of E10 to BD30 and the ID was shortened by reducing the alcohol fraction. In the previous study, the engine used was a type T40 M, D-144 diesel engine, which was newly acquired for research in 2003 and was used throughout the period 2004–2008 [22]. Further details of the engine include the following: (a) mechanical power of 37 kW on 1600 rpm, (b) naturally-aspirated; (c) direct-injection, (d) four-cylinders; (e) diameter of 110 mm; (f) stroke of 120 mm, (g) compression ratio was 16.5:1, (h) dished-piston; and (i) compression ignition engine. The fuel was delivered by a single-plunger, fuel-pump, and type: ND 21/4, through three-hole injection-nozzles; with initial fuel delivery starting at 25 $\pm$ 1° BTDC. The needle-valve lifting pressure on the injectors was set to 17.5 $\pm$ 0.5 MPa.

5.1. Analysis

Both biodiesel and n-butanol fractions in DF reduces the combustion duration of the blend. However, n-butanol has a limitation and may not be increased any further than 20% (v/v) in

DF due to its significantly low CN. Biodiesel, on the other hand, has a higher CN making it possible to increase fractions for blending in DF. Results obtained in the author's study involving only n-butanol in DF and comparing with a study by others [20] who included biodiesel in n-butanol/DF blend indicated enhanced combustion characteristics when biodiesel was added. It was observed that B50 performed better than Bu20.

Figure 6 illustrates the standard deviation of pressure cycles. It can be deduced that blends have less deviation of pressure from the mean value than diesel. In other words, blends have a more stable combustion characteristic than the reference fuel. The blend B10 revealed a less stable combustion quality than the other blends by indicating a higher standard deviation at all speeds. A similar study by others [20] measured the thermodynamic cycle-to-cycle variations for the in-cylinder pressure using the coefficient of variation (COV) to determine combustion stability for the blends Bu20 and B50. They found that the COV for the blends was below 5% for the engine loads, which agrees with the author's study where the standard deviation of the mean in-cylinder pressure cycles was below that of DF. **Figure 6** results should reconsider referring to other studies.

6. Soot concentration from (TDI) compression ignition engine

Figure 7 illustrates the soot concentration in the exhaust gas at 1500 and 3000 rpm. In a direct-injection diesel engine, the fuel-air distribution is not homogeneous. Therefore, in addition to the excess-air ratio, soot formation is also controlled by the mixing of air and fuel [19]. The formation of soot depends upon the condition of the engine, the type of the combustion chamber and differences in the physical and chemical properties of the fuel [19]. The soot emission is reduced when a shared volume of n-butanol is added to DF. This is because the oxygen atoms attached to the hydroxyl group of n-butanol hinder (through moderation) the production of the precursors responsible for the formation of soot [21]. The low C/H ratio of n-butanol also reduces soot formation. In Study P, as well as in the author's study, the soot concentration was substantially reduced. The soot concentration reduced when n-butanol/DF

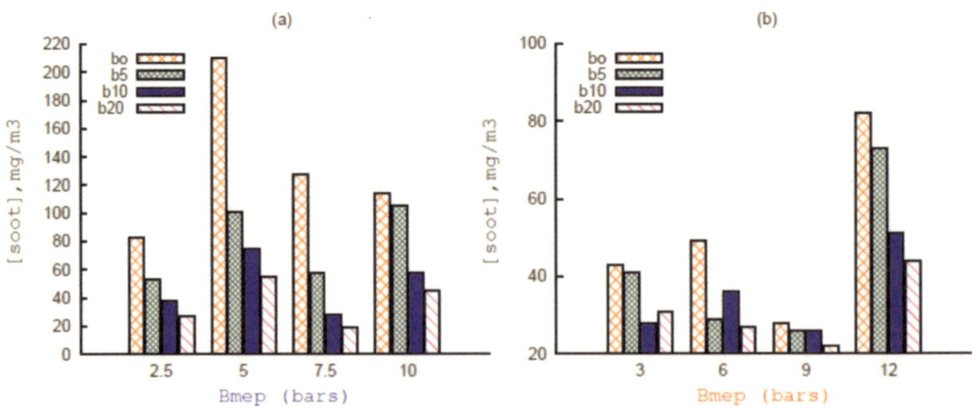

Figure 7. Soot emissions against varying BMEP (a) at 1500 rpm (b) at 3000 rpm [21].

blends were used as indicated in **Figure 7**. The soot concentration for the blends was always lower than that of DF in both the speeds: 1500 and 3000 rpm, when measured against BMEP. The soot emission reduction with increasing the shared volume of n-butanol to DF was 55.5, 77.8, and 85.1% for B5, B10, and B20, respectively in the 75% load with 1500 rpm. The reduction of soot emission for all the test fuels was higher in the engine tests with 3000 rpm than with 1500 rpm. The reduction of soot emission was highest in the 75% load. The small deviation exhibited by the blends from the trend in the 25 and 50% load with 3000 rpm is not well known: it might be caused by the temperature distribution during the combustion process, as temperature also plays an important role in the formation and oxidation of soot.

7. Conclusions

A study was conducted to determine the performance and combustion characteristics of a mixture of n-butanol in diesel fuel in the ratios: 5,10, and 20% (B5:n-butanol 5% with diesel fuel 95%, B10 and B20) using a turbo-charged four cylinder compression ignition engine. The results were compared with studies by others using similar shared volumes of n-butanol (20%, v/v), including 40% (v/v) biodiesel. Combustion characteristics of B20 (n-butanol 20% and 80% DF) in the author's study improved when the study was compared with the study by others where 40% biodiesel was added to B20. A higher value of the standard deviation for DF than the blends was observed from the standard deviation diagram, indicating a more stable combustion process for the blends than DF. Soot emission was greatly reduced in both compared studies (in author's study and Ref. [19]). This was when bioethanol admix was introduced to BD30/DF blend in the other study [19]. In both cases, smaller proportions of bioalcohol were used. In the author's study, the soot reduction relative to DF at 1500 rpm at 75% load for B05, B10, and B20 mixtures was 55.5, 77.8, and 85.1% respectively. This reduction is a significant advantage of blending DF with smaller shared volumes of bioalcohol. The study has indicated a highly prospective fuel in n-butanol/DF bioalcohol to be promoted in the blending science to reduce particulate matter and improve combustion efficiency in the application of the diesel fuel in reciprocating internal combustion engines.

Moreover, the results should be compatible with figures. I cannot understand how authors say, for example, about ID using pressure development figures, premixed combustion period using HRR or combustion stability which requires a cyclic analysis. Certainly, pressure development and HRR curves are a indicator for such parameters, but the author should define, first, the terms based on the related quantity for better understanding of a reader. Moreover, the results should be discussed more.

Acknowledgements

The authors acknowledge and are greatly indebted for the financial support from the joint research collaboration between Hungary/South Africa Funding (UID 72384 and TET_10-1-2011-0005); for facilitation by the two universities, Tshwane University of Technology, Pretoria,

South Africa and Budapest University of Technology and Economics, Budapest, Hungary; the latter for making available the laboratory facility, Jendrassik Gyorgy hotechnikai Laboratorium in Budapest BME, where the engine experiments were conducted. This work is connected to the scientific program of the "Development of quality-oriented and harmonized R+D+I strategy and functional model at BME" Project. This Project was supported by the New Szechenyi Development Plan (Project ID:TAMOP-4.2.1/B-09/1/KMR-2010-0002). The authors are grateful to the support of Bolyai Janos Research fellowship of HAS (Hungarian Academy of Science). Authors are grateful for the support of Prof. Dr. Florian Heinitz, Director of Transport, and Spatial Planning Institute in the University of Applied Science in Erfurt, Germany.

Nomenclature

ATDC	After top dead center
CA	Crank angle
BSFC	Brake-specific fuel consumption
ID	Ignition delay
CAD	Crank angle degree or θ crank angle
EGT	Exhaust gases temperature
BTDC	Before top dead center
HRR	Heat release rate
rpm	Revolution per minute
NOx	Nitrogen oxides
BTE	Brake thermal efficiency
CO	Carbon monoxide
BMEP	Brake mean effective pressure
CN	Cetane number
DF	Diesel fuel
IP	Indicated pressure
EDC	Electronic diesel control
UHC	Unburned hydrocarbon
MAP	Manifold air pressure

Greek letter

λ Excess air ratio

Author details

Lennox Siwale[1]*, Lukacs Kristof[2], Torok Adam[2], Akos Bereczky[2], Makame Mbarawa[3], Antal Penninger[2] and Andrei Kolesnikov[4]

*Address all correspondence to: zumbe.siw@gmail.com

1 Department of Mechanical Engineering, Copperbelt University, Kitwe, Zambia

2 Department of Energy Engineering, Budapest University of Technology and Economics, Budapest, Bertalan Lajos, Hungary

3 Ministry of Communication, Science and Technology, Dar es Salaam, United Republic of Tanzania

4 Department of Mechanical Engineering, Tshwane University of Technology, Pretoria, South Africa

References

[1] Demain A. Biosolutions to the energy problem. Industrial Microbiology and Biotechnology. 2009;**36**:319-332

[2] Nelson DK, Lapara TM, Norak PJ. Effects of ethanol-based fuel contamination: microbial community of regulated compounds, and methane generation. Environmental Science & Technology. 2010;**44**:4525-4530

[3] Demirbas A. Biofuels securing the planet's future energy needs. Energy Conversion and Management. 2009;**50**(9):2239-2249

[4] Mustafa B. Production of bioethanol from lignocellulosic materials via the biochemical pathway: A review. Energy Conversion and Management. 2011;**52**(2):858-875

[5] Dogan O. The influence of n-butanol/diesel fuel blends utilization on a small diesel engine performance and emissions. Fuel. 2011;**90**(7):2467-2472

[6] Zhang Y, Boehman AL. Oxidation of 1-butanol and a mixture of n-heptane/1-butanol in a motored engine. Combustion and Flame. 2010;**157**:1816-1824

[7] Zhang Y, Boehman AL. Oxidation of 1-butanol and a mixture of n-heptane/1-butanol in a motored engine. Combustion and Flame;**157**(10):1816-1824

[8] Yao M et al. Experimental study of n-butanol additive and multi-injection on HD diesel engine performance and emissions. Fuel. 2010;**89**(9):2191-2201

[9] Surisetty V et al. Alcohols as alternative fuels: An overview. Applied Catalysis A: General. 2011;**404**(1-2):1-11

[10] Sarathy SM et al. An experimental and kinetic modeling study of n-butanol combustion. Combustion and Flame. 2009;**156**(4):852-864

[11] Hsieh W-D et al. Engine performance and pollutant emission of an SI engine using ethanol–gasoline blended fuels. Atmospheric Environment. 2002;**36**(3):403-410

[12] Rakopoulos DC et al. Effects of butanol-diesel fuel blends on the performance and emissions of a high-speed DI diesel engine. Energy Conversion and Management. 2010;**51**(10):1989-1997

[13] Bryan W et al. Autoignition of n-butanol at elevated pressure and low-to-intermediate temperature. Combustion and Flame. 2011;**158**:809-819

[14] Karabektas M, Hosoz M. Performance and emission characteristics of a diesel engine using isobutanol-diesel fuel blends. Renewable Energy. 2009;**34**:1554-1559

[15] Kumar S et al. Advances in diesel-alcohol blends and their effects on the performance and emissions of diesel engines. Renewable and Sustainable Energy Reviews. 2013;**22**:46-72

[16] Broustail G et al. Experimental determination of laminar burning velocity for butanol and ethanol iso-octane blends. Fuel. 2011;**90**(1):1-6

[17] Laza T et al. Examination of burning processes regenerative liquid fuel and alcohol mixtures in diesel engine. Periodica Polytechnica Mechanical Engineering. 2006;**50**:11-26

[18] Sayin C. Engine performance and exhaust gas emissions of methanol and ethanolol-diesel blends. Fuel. 2010;**89**(11):3410-3415

[19] Raslavicius L, Bazaras Z. Variations in oxygenated blend composition to meet energy and combustion characteristics very similar to the diesel fuel. Fuel Processing Technology. 2010;**91**(9):1049-1054

[20] Ibrahim A. Performance and combustion characteristics of a diesel engine fuelled by butanol-biodiesel-diesel blends. Applied Thermal Engineering. 2016;**103**(Supplement C):651-659

[21] Siwale L et al. Combustion and emission characteristics of n-butanol/diesel fuel blend in a turbo-charged compression ignition engine. Fuel. 2013;**107**(2013):409-418

[22] Siwale L. (zumbe.siw@gmail.com) age of the engine used for research [email-to-], 2013 (Raslavicius L, (Laurencas.Raslavicius@gmail.com), Dept of transport engineering at kaunas, University of technology Luthiana, July 16)

Chapter 4

Reactivity Controlled Compression Ignition (RCCI) of Gasoline-CNG Mixtures

Firmansyah, Abdul Rashid Abdul Aziz,
Morgan Raymond Heikal,
Ezrann Zharif Zainal Abidin and
Naveenchandran Panchatcharam

Additional information is available at the end of the chapter

http://dx.doi.org/10.5772/intechopen.72880

Abstract

Reactivity controlled compression ignition (RCCI) is a dual fuel combustion method that relies on the significant difference in reactivity of the fuels involved. RCCI had a low performance at high engine speed due to its high tendency on knocking and high pressure rise rate. Therefore, this study investigates the effect of the fuel stratification on the RCCI combustion and its extended to the interaction of two low reactive fuels, gasoline and compressed natural gas (CNG), in the RCCI combustion system. The investigation was experimentally performed on a single cylinder engine and constant volume chamber. The stratification was created by varying injection timing in the engine by injecting CNG at 80° and 120° before top dead center (BTDC) and varying injection gap in the constant volume chamber with the gaps between two fuel injection timing were varied between 0 ms to 20 ms. The results in the engine experiment show that proportions of gasoline and CNG and degree of stratification of CNG were found to be effective means of combustion control within certain limits of engine load and HC and CO emissions could be significantly reduced. While in constant volume chamber it has a significant effect on the combustion phasing. Stratified mixture produces shorter combustion duration while homogeneous mixture produces longer duration.

Keywords: reactivity charge compression ignition, gasoline, compressed natural gas, stratification, fuel injection

1. Background

Perhaps the most graceful invention by humankind that ever had a greater impact on society, the economy, and the environment is the reciprocating internal combustion engine, in general, called the IC engine. For decades, this magnificent invention proved to play a vital role in the automobile system, used almost exclusively today. There are two types of internal combustion engines: spark ignition (SI) and compression ignition (CI). For the last decades, rapid improvements in the efficiency have been achieved on both types of IC engine.

Unfortunately, at present, there is a pressing need to develop advanced combustion engines that maximize the engine efficiency and totally mitigate the exhaust pollutants. Profound understanding of both SI and CI combustion principles has been achieved during the last decades to improve the efficiency and reduce the emissions. The conventional SI combustion, which is characterized by flame propagation in near-stoichiometric homogeneous mixtures, produces very low exhaust emissions in combination with a three-way catalytic converter but has a relatively low thermal efficiency, which is its main drawback.

CI combustion, on the other hand, that is characterized by the autoignition of a lean fuel-air mixtures, has a very high thermal efficiency; yet, it has very high soot and NO_x (nitrogen oxide) emissions. Diesel engines typically produce lower carbon monoxide (CO) and unburned or partially burned hydrocarbons (HC) compared to the gasoline engines. However, NO_x, which comprises nitric oxides and nitrogen dioxides, in addition to particulate matter (PM) or soot, is significant pollutant from diesel engines, which require proper control strategies as they pose adverse health and environmental impacts.

The engine technologies are advancing at a significant rate during the last decades. The engine technology development timeline is depicted in **Figure 1**. The engine performance was the main priority in the first era of engine developments. Technologies such as turbocharger, port fuel injection, high compression ratio, direct fuel injection, and engine lightweight material were the technologies that are focusing on increasing the engine power output to its maximum capabilities. This early era was driven by the abundant amount of fuels relative to its low demand as the automobile was still an exclusive technology.

The second era of the engine technology development was mainly driven by increasing concern about exhaust emissions and efforts in achieving low fuel consumption. In the earlier technologies, efficiency was improved and the engine downsizing was the primary target of the engine development. Homogeneous charge compression ignition (HCCI) engine is one of the promising alternatives in order to achieve these objectives.

The spark ignition engine (SI) and compression ignition engine (CI) are the established engine technologies, and each have their advantages and disadvantages. SI has a faster response and low emissions yet low efficiency, while CI offers high efficiency and low fuel consumption yet higher emissions and slower response. This makes the development of the SI and CI engine followed different approaches. Nevertheless, the main objective of an engine is mainly to achieve high performance, high efficiency with low emissions. In order to achieve this goal,

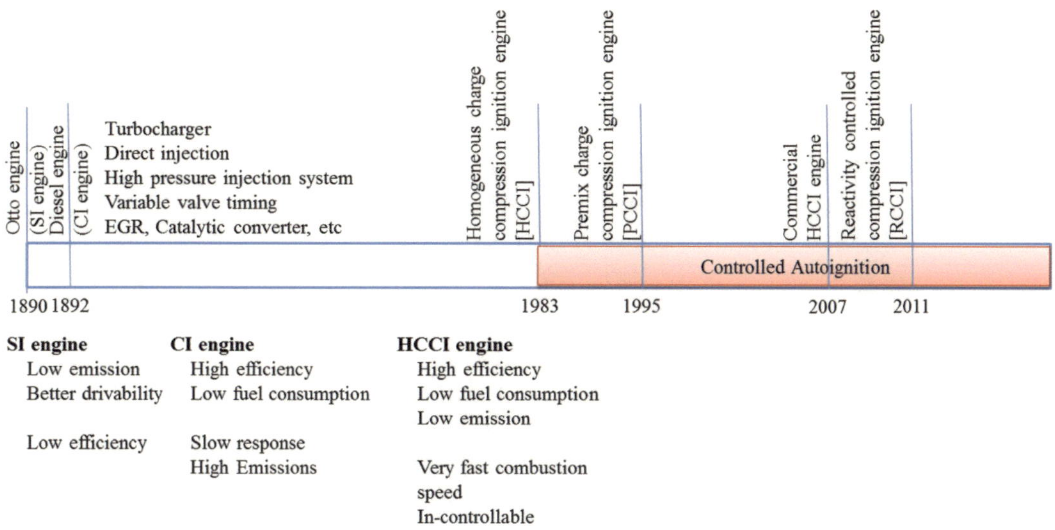

Figure 1. Engine technology development.

HCCI was introduced in 1983. Major obstacle in HCCI is its rapid heat release rate with a very high maximum pressure, which is detrimental to engine structure. In the effort of controlling the combustion in HCCI engines, technologies such as exhaust gas recirculation (EGR) and variable valve timing (VVT) are utilized to improve the controllability of HCCI combustion, which is used in the commercial HCCI engine such as skyActive technology (Mazda) and diesOtto engine (Mercedes).

On the other hand, in an effort to find the best method for controlling HCCI combustion, the HCCI has evolved into various types of controlled autoignition-based engine (CAI) such as premixed charge compression ignition (PCCI) in 1995 [1] and reactivity controlled compression ignition (RCCI) engine in 2011 [2]. These methods are proven to be able to improve the operating range of CAI engine. However, there are no established methods that are proven to be effective in controlling CAI engine.

Figure 2 depicts the differences in each combustion system control strategy. The CI system (**Figure 2** (**a**)) creates a very high pressure inside the combustion chamber, and the fuel is directly injected to combust. SI system (**Figure 2b**)), on the other hand, forms a homogeneous mixture either by direct injection or by port injection before the mixture is ignited by a spark, whereas the HCCI engine (**Figure 2** (**c**)) produces a homogeneous mixture by injecting the fuel during the intake stroke, and the mixture is autoignited due to the compression. HCCI combustion combines the best features of gasoline and diesel engines to produce diesel-like power and efficiency while maintaining gasoline-like soot free emissions within certain operating limits.

The engine performance was limited to the part load conditions and controlling the combustion process was very problematic due to the autoignition being highly dependent on the

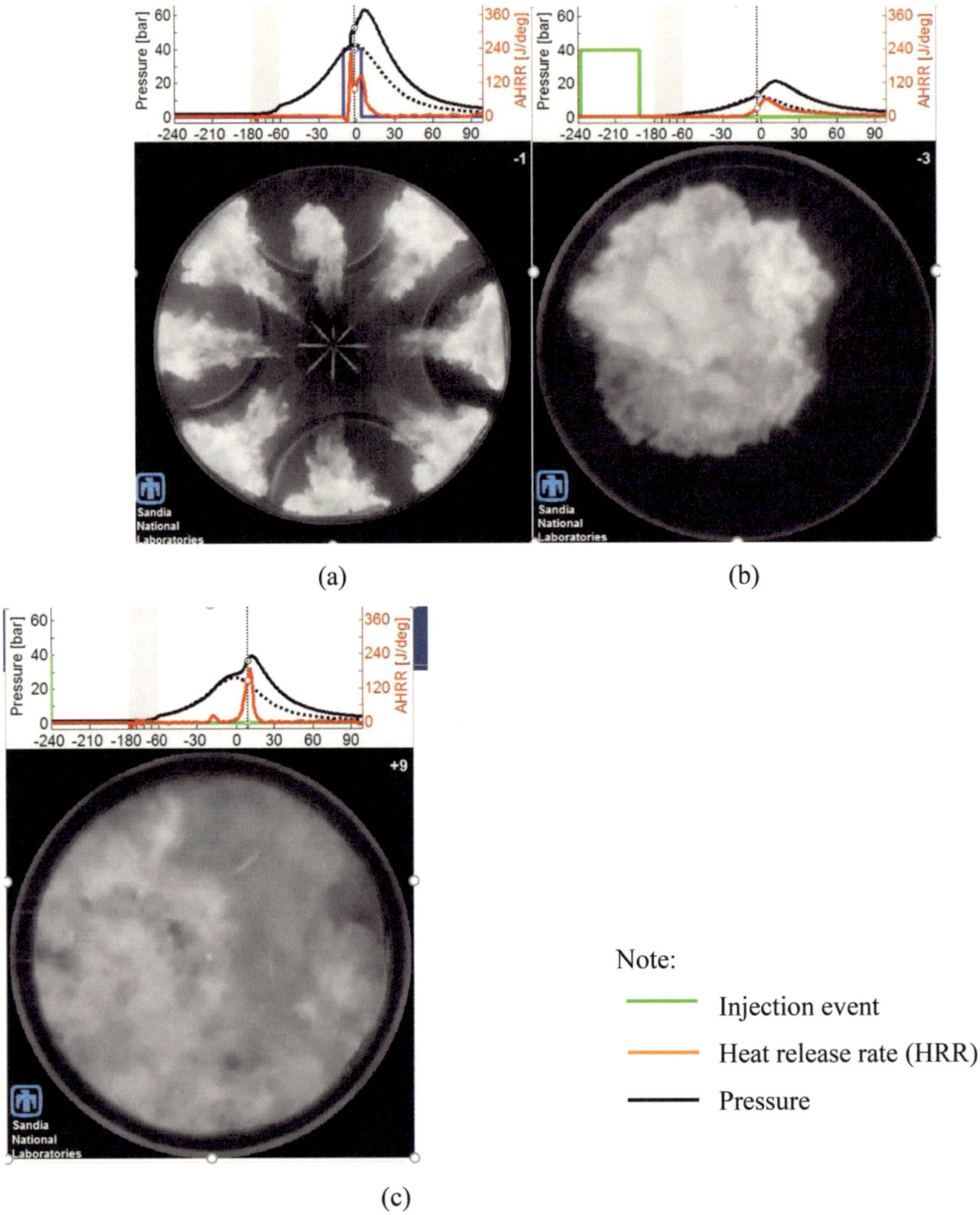

Figure 2. Combustion control strategies (a) compression ignition (CI), (b) spark ignition (SI), and (c) homogeneous charge compression ignition (HCCI). Source: http://crf.sandia.gov/combustion-research-facility/engine-combustion/fuels/.

temperature, pressure, and mixture composition inside the combustion chamber. In its development, HCCI was limited by the narrow operating range and the unpredictable combustion delay and behavior. The maximum pressure generated was very high but produced a low mean effective pressure due to the short combustion duration. Many methods and possibilities were proposed to control HCCI engines. In this process, the paradigm of creating an autoignition process from homogeneous charge is shifting to the method of controlling autoignition. The first development is the premixed charge compression ignition (PCCI). This concept was introduced by Aoyama et al. [1]. Gasoline was subjected into diesel-like environment with high compression ratio 17.4:1. A port injection method was used to create the premixed charge in the combustion chamber where the fuel is injected very close to the intake valve closing time as shown in **Figure 3**.

The combination starts with gasoline and diesel as the low-reactive fuel and high-reactive fuel, respectively. It was found that RCCI combustion was able to operate in a wide range of engine loads with near-zero NO_x and soot emissions, accepted pressure rise rate and high indicated efficiency. However, RCCI still could not achieve high load operation with power outputs comparable to CI engine. The combustion behavior of RCCI is somewhat still unpredictable. Further investigation on the important parameters in RCCI combustion control will improve the understanding of the combustion process of RCCI that leads to better control of the process and better engine output.

The premixed charge compression ignition (PCCI), reactivity charge compression ignition (RCCI), and spark-assisted HCCI are some of the established methods of controlling the autoignition process in an engine. All of these methods are categorized as CAI engines. Regardless of the limitations of the controlled autoignition (CAI)-based combustion system, CAI offers high efficiency [4], low fuel consumption [5], and low emission [6], which are the main aims of future engine development. The attributes that differentiate SI, CI, HCCI, PCCI, and RCCI are shown in **Table 1**.

These earlier works are the basis of the controlled autoignition engine concept. The primary focus of the CAI combustion concepts is identifying the relevant influencing parameters as well as control parameters of this system to widen the operating range and improve the efficiencies. The need for a thorough understanding of the CAI combustion process initiates further discussion on the method to control its combustion. The next stage is to determine the engine parameters that have a direct effect on the combustion. As these steps are carefully defined, high efficiency, low fuel consumption, and low emissions internal combustion engines are achievable (**Table 2**).

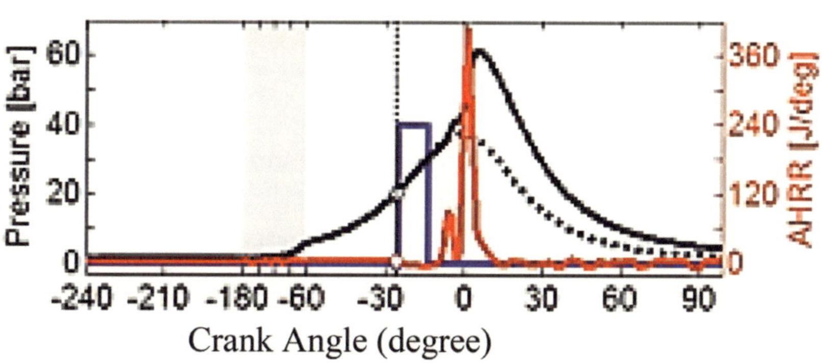

Figure 3. PCCI combustion method [3].

Many researches have been done in the effort of controlling CAI combustion-based system. Agarwal et al. [7] summarize the various types of combustion method and some method in controlling the combustion process. There are various areas that require improvement in order to reshape the combustion and improve the efficiency while still having a low emission. Development of control on ignition timing [8], method in slowing down the heat release rate at high load [9, 10] and development of intake and exhaust manifold for multicylinder engine are among the few area of improvement that have been identified in the area of CAI combustion system. Focusing on the RCCI combustion control method, Li et al. [11] categorized the control by two main categories, fuel and engine management. The fuel management includes two fuel strategies [12, 13] and single fuel strategy with additives, while the engine management

Combustion mode	SI	CI	HCCI	PCCI	RCCI
Fuel	Gasoline-like fuels	Diesel-like fuels	Flexible fuels	Diesel-like fuels	Multifuels
Lambda	1	1.2–2.2	>1	>1	>1
Mixture preparation	PFI, GDI	DI	DI, PFI and DI + PFI	DI, PFI	DI, PFI and DI + PFI
Ignition	Spark ignition	Autoignition	Autoignition	Autoignition	Autoignition
Combustion form	Premixed	Diffusion	Premixed but dominated by chemical kinetics	Premixed	Premixed + stratified
Combustion rate limitation	Flame propagation	Mixing rate	Multipoint or spontaneous	Multipoint or spontaneous	Fuel reactivity
Flame front	Y	Y	w/o	w/o	Y
Combustion temperature	High	Partially high	Relatively low	Relatively low	Relatively high

Table 1. Traditional and the controlled autoignition (CAI)-based combustion mode.

Fuel type	Gaseous fuel
Fuel supply system	Direct injection (DI) **a.** High-pressure direct injection (12–18 bars) **b.** Air-assisted low pressure injection (4–6 bars)
No. of cylinders bore Stroke compression ratio	Single cylinder (399.25 cc) 88 mm 132 mm 14:1 (Geometric)
No. of valves Valve timing events Intake valve open (IVO) intake Valve closure (IVC) Exhaust valve open (EVO) Exhaust valve closing (EVC)	4 12° BTDC 132° BTDC 15° BBDC 10° BBDC

Table 2. Summary of specifications of the engine.

include fuel ratio [14], injection strategy [15–18], EGR rate [19], compression ratio [16], bowl geometry [20, 21], stability control, and utilization of two injectors.

This chapter introduces two low reactive fuels, gasoline and CNG, in an RCCI combustion system in order to increase the limit of RCCI engine operation. It introduces a method that has different principals compared to RCCI by introducing combination of low reactive fuels rather than combination of high- and low-reactive fuel. Furthermore, this chapter also introduces the method of RCCI combustion control by varying the stratification by varying the injection timing and gap between two fuel injectors. Two approaches were done to investigate the behavior of gasoline-CNG mixtures in the RCCI combustion system. First approach is experimental testing on a single cylinder engine that was converted to dual fuel engine with gasoline injected at intake port, while CNG is directly injected into the combustion chamber. The stratification was done by varying the injection timing of CNG, while gasoline is kept homogeneous.

The second approach is the combustion testing in a constant volume chamber with both fuels that are directly injected into the combustion chamber. In this setup, the stratifications level in chamber is done with varying the injection gap between two fuel injections.

2. Test procedure and equipment

The test procedure and equipment used in both approaches are elaborated in this section.

2.1. Engine testing and equipment

The engine used for this experimental study houses the fuel system of direct injection of gaseous fuel and has compression ratio of 14. This engine is a single cylinder water-cooled engine coupled to an electric dynamometer that can be used for starting the engine and measuring the brake torque produced by the engine. **Figure 4** shows a schematic drawing of the engine.

An electric heater is provided to heat the lubricant oil to help warm up the engine faster. A separate control unit controls the operation of the pumps and the temperature of the oil and water by temperature controllers. The control unit also controls the operation of the dynamometer, which also serves as the starter motor and the engine can be motored at a wide range of speeds. There are standard features of safety included in the control unit such as emergency switch, automatic shut down upon the excessive rise in the oil and/or coolant temperatures, or any abnormal conditions of electrical power supply (**Figure 5**).

A commercially available gasoline port fuel injector was used and its specifications can be found in **Table 3**. This injector has low flow rates and was selected to match the requirement of injecting very low quantities of gasoline to operate in HCCI mode with ultralean mixtures. The injector comes calibrated in the factory to inject and precisely meter the volume against the specified injection duration.

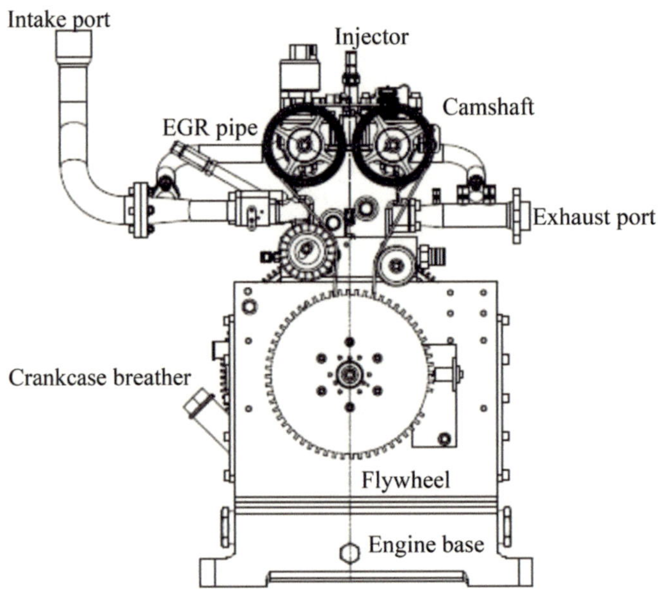

Figure 4. The single cylinder research engine.

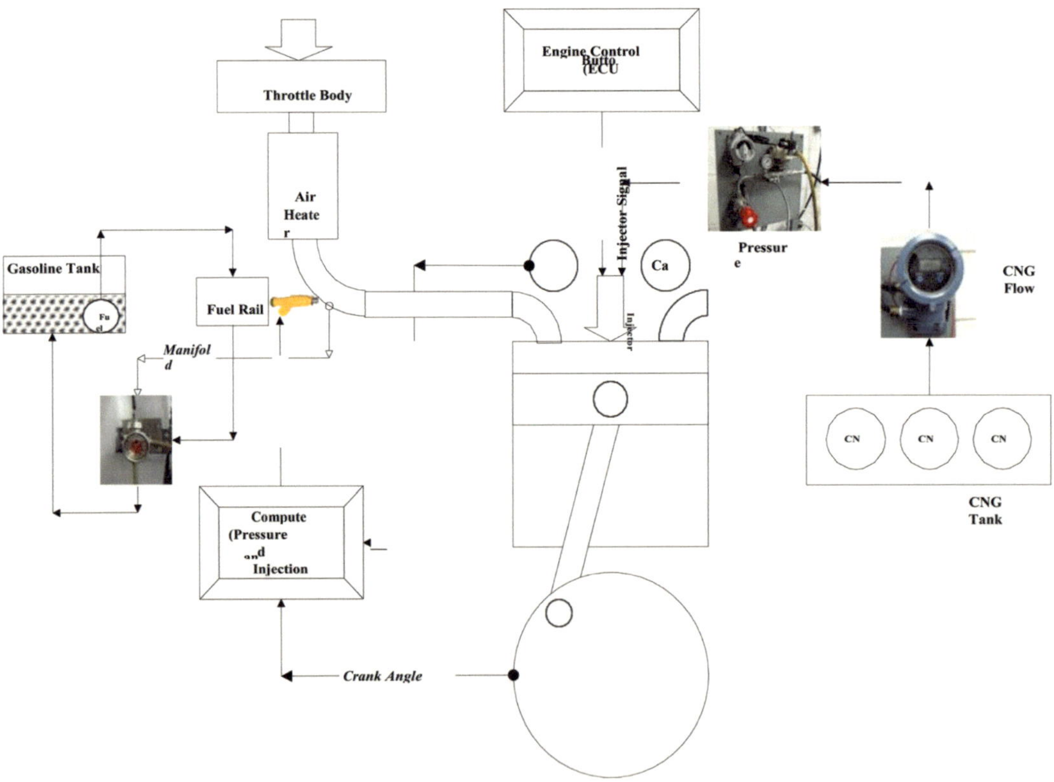

Figure 5. Gasoline and CNG fuel supply systems.

http://dx.doi.org/10.5772/intechopen.72880

Make/part number	Bosch/0280155710	
Type	High impedance	
Fuel injection pressure	3 bar	
Fuel flow rate	191.8 cc/min	
Power supply	12 V DC	

Table 3. Specifications of the gasoline injector.

Engine speed (N)	Equivalence ratio of gasoline (φg)	Timing of CNG injection (SOI)	CNG quantity (mCNG)
RPM	**Ratio**	**CAD-BTDC**	**mg/cycle**
1200	0.20	300	Up to the knocking limit or unstable engine operation
1500	0.22	240	
1800	0.24	180	
2100	0.26	120	
		80	

Table 4. Matrix of experiments.

Fuel compatibility	Standard gasoline and ethanol flex fuels
Fuel pressure	300 kPa
Static flow rate at 300 kPa	7.55 g/s
Offset	0.67 ms
Gain	0.11 = 9.09 mg/ms
Minimum linear PW	1.5 ms
Linear flow range	8.43
Open/closing time	1.3/0.7 ms
Coil resistance	12
Injector inductance	11.6 mH
SCOV/DMOV	4.42/4.92 V
Spray pattern	26°Cone

Table 5. Injector technical specification.

The change in air flow rate and volumetric efficiency at different engine speeds were taken into account, and the gasoline injection durations were adjusted so as to operate with constant equivalence ratios of gasoline from 0.20 to 0.26. **Table 4** shows the matrix of experiments and lists the variable parameters of the experiments conducted for this study.

The CNG quantity was varied and limited by the combustion stability of the mixture by limiting the maximum coefficient variation (CoV) of the combustion to 10%. Thus, the readings were obtained at different speeds, different gasoline flow rates, and different injection timings with various quantities of CNG injected (**Table 5** and **Figure 6**).

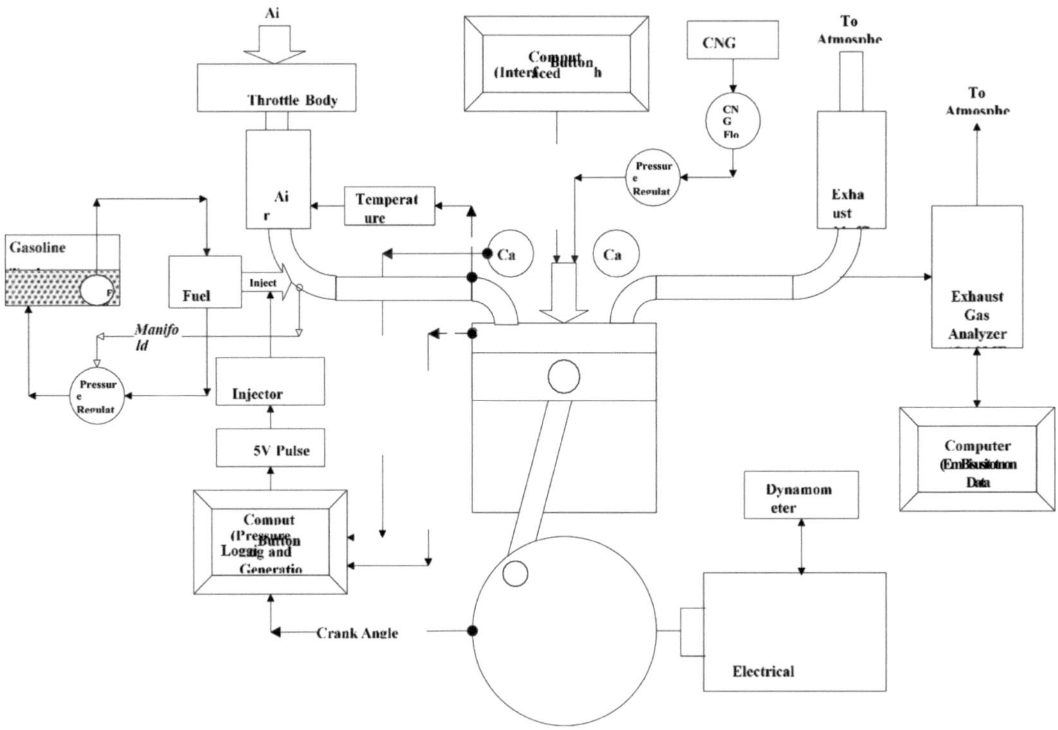

Figure 6. Experimental setup.

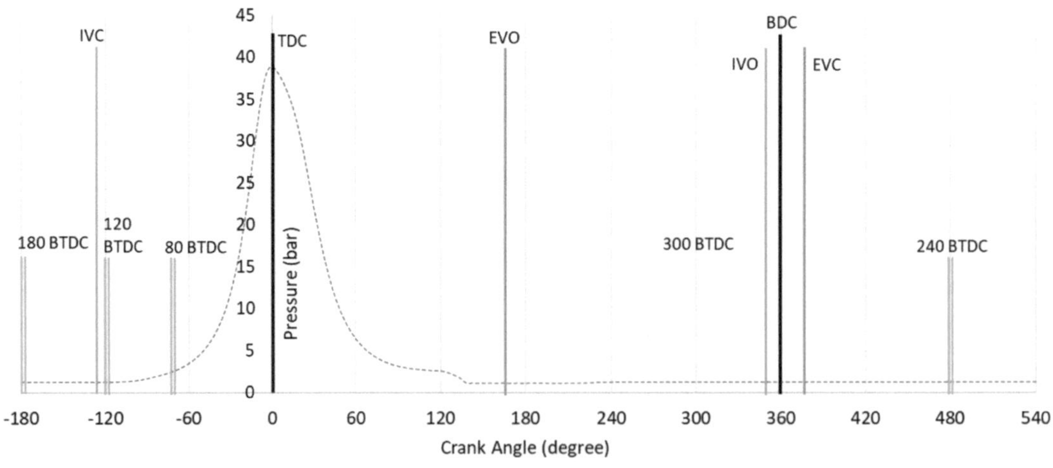

Figure 7. Injection timing of CNG, and intake and exhaust valve timing.

Figure 7 depicted the injection timing, and intake and exhaust valve timing of the engine. It shows that injection timing 300, 240, and 180° BTDC take place before the intake valve close, while 120 and 80 BTDC take place after intake valve close.

2.2. Constant volume combustion chamber testing and equipment

Following section describes the detailed experimental setup and equipment used on the constant volume chamber. Constant volume combustion was used as the primary method for characterizing the interaction of the parameters such as fuel compositions, lambda, and mixing ratio. The equipment and control system were designed, manufactured, and calibrated to accommodate the experimental works. The primary data in this experimental setup are combustion images and pressure trace of which acquired by Schlieren and direct measurement method, respectively.

The experiments were carried out in a constant volume combustion chamber with diameter and length 80 and 100 mm, respectively. A 100 W cartridge heater was placed in the middle of the chamber to increase in-cylinder gas temperature with a maximum temperature of 820 °C. There were two window access planes to the chamber in order to facilitate the Schlieren image visualization method (**Figures 8–10**).

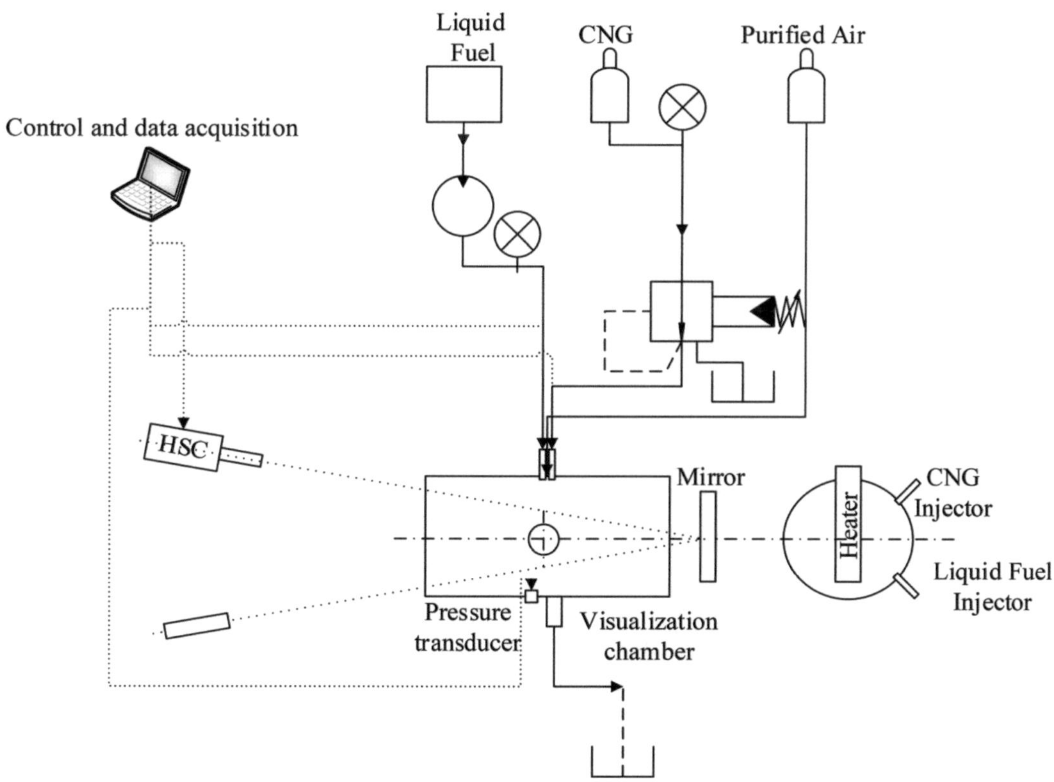

Figure 8. Experimental setup.

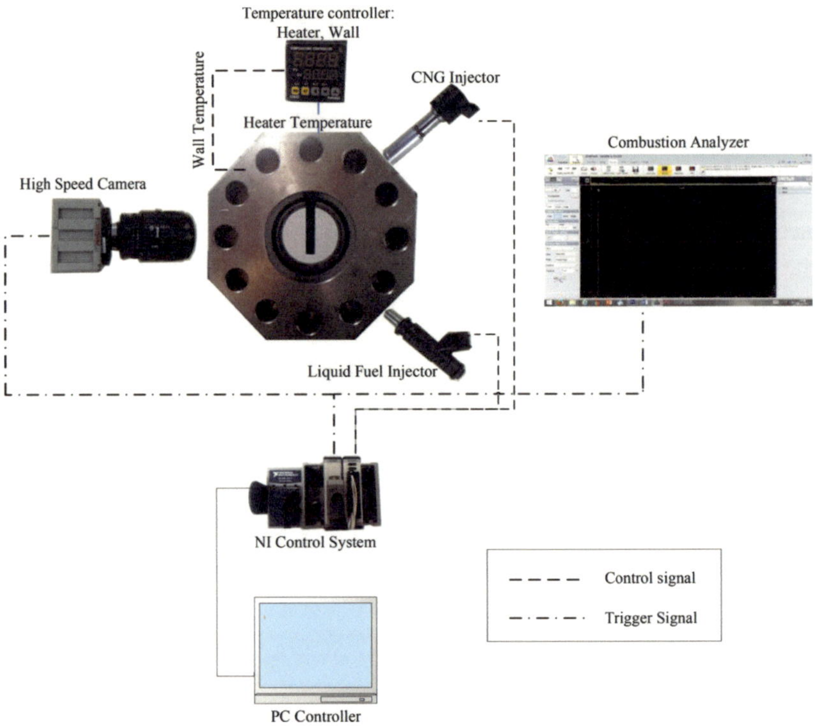

Figure 9. Schematic diagram of the control system and data acquisition.

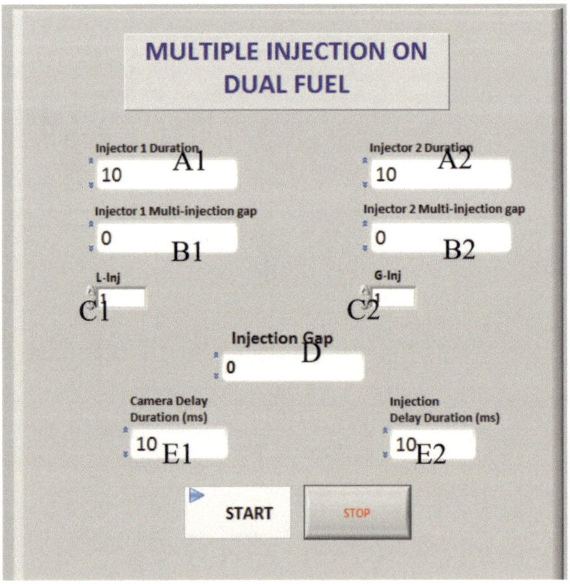

Figure 10. Injector control system interface.

3. Gasoline-CNG mixtures in RCCI combustion system

The gasoline-CNG mixtures performance and combustion in RCCI combustion system from both methods are elaborated in this section. It explained the parameters affecting the combustion of low-reactive fuels and method in controlling the combustion process.

3.1. Gasoline-CNG combustion behavior in RCCI combustion-based engine

The degree of stratification of CNG in the total mixture was found to have significant effects on the maximum load in terms of the IMEP attainable and φTotal. The degree of stratification is determined by the injection timing with 300° BTDC representing the homogeneous mixture, while 120° BTDC represents the stratified mixture. The 300° BTDC has a very early injection timing, and the fuel is injected during intake valve is open. It is, therefore, the fuel that has sufficient time to be completely mixed with air and create a homogeneous mixture. While on the other hand, for 120° BTDC, the fuel is injected after intake valve closed and mixing time between fuel and air is very short and does not allow a complete mixing to take place.

Figure 11 shows that at 300 and 240° BTDC injection higher total equivalence ratios could be operated. But with 180° and 120° BTDC, the maximum operable φTotal was limited with reduced IMEP at a given φTotal when compared to the 300 and 240° BTDC cases.

The IMEP results show agreement with investigation from Genchi G and Pipitone E [22] where the increased composition of CNG produces higher IMEP. With the highest degree of stratification, although the maximum load was limited, there was no significant drop in the IMEP and the trend was similar to 300 and 240° BTDC conditions. The corresponding values of indicated thermal efficiencies are shown in **Figure 12**. The maximum load was observed to be limited by knocking when CNG was injected at 300° BTDC, and for the other cases, increasing CNG injection rate led to unstable operation or misfire.

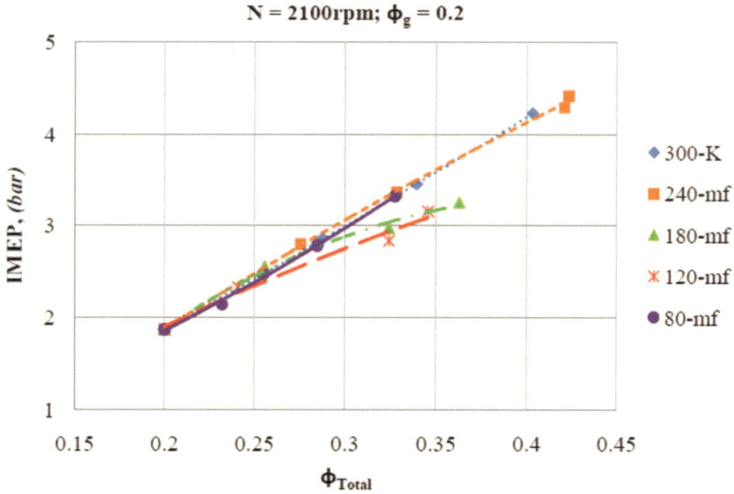

Figure 11. Effect of degree of stratification of CNG on IMEP. (K—limited by knocking; mf—limited by misfire).

From **Figure 13**, it can be seen that the ignition timing could be altered by changing the timing of injection of CNG at a given load. The ignition timing was determined by identifying the start of heat released rate and mass fraction burned derived from the pressure data where the 0% points before the continuous propagation of the mass fraction burned is determined as the start of ignition of the analyzed combustion cycles.

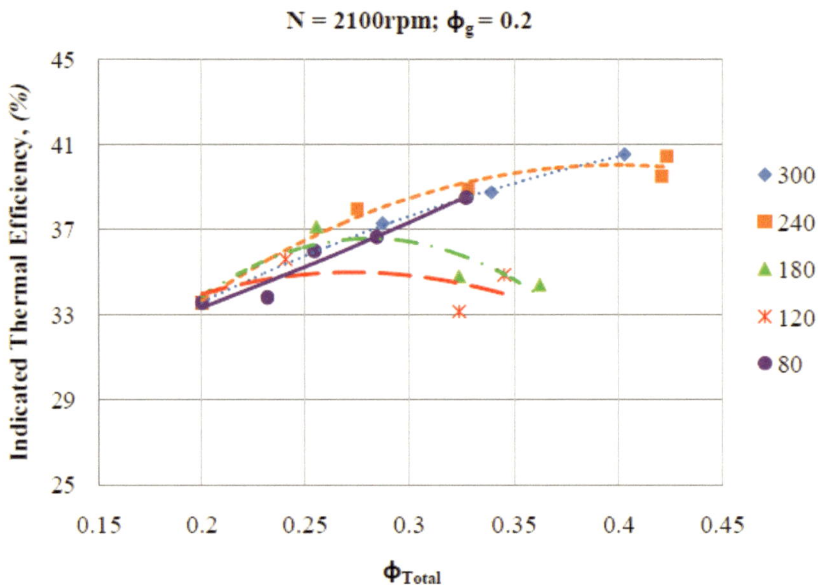

Figure 12. Effect degree of CNG stratification on the indicated thermal efficiency.

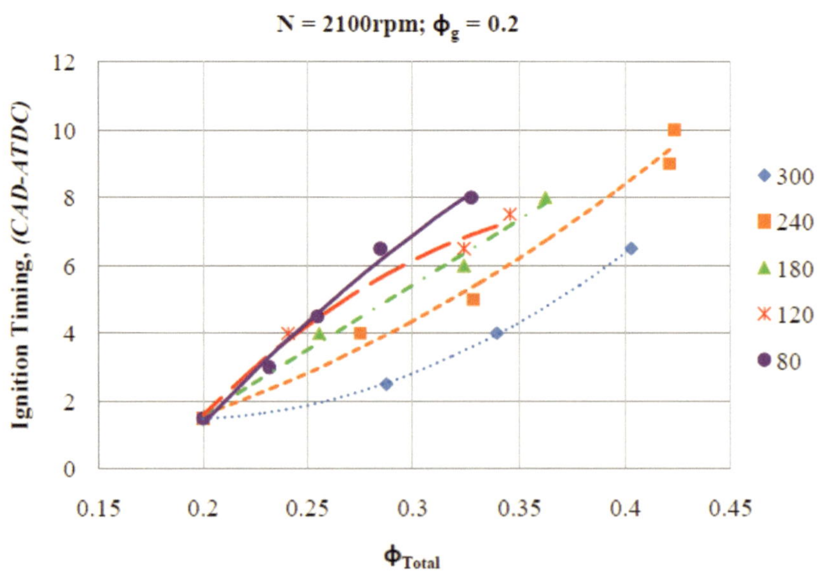

Figure 13. Effect of degree of CNG stratification on ignition timing.

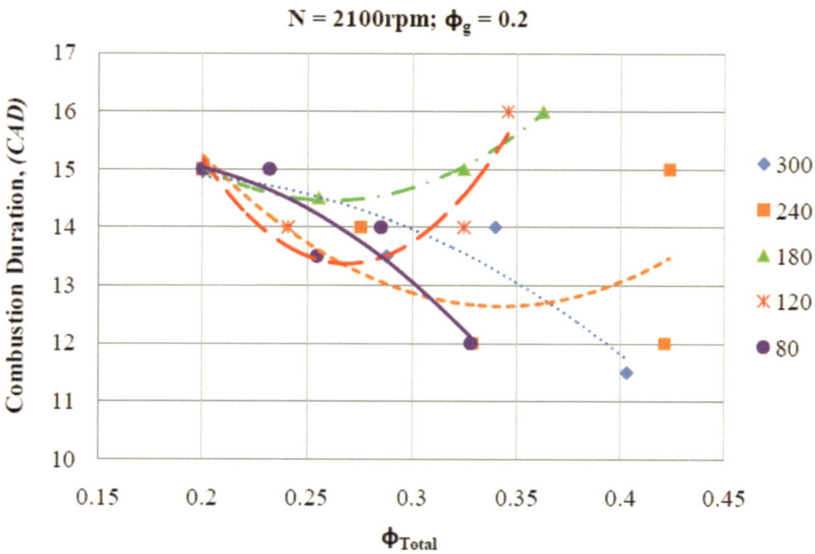

Figure 14. Effect of degree of CNG stratification of CNG on the combustion duration.

When the rate of CNG injection was increased, the ignition timing was delayed due to the higher octane number of CNG. Also, higher degrees of stratification resulted in higher increments in the delay of ignition timing as the CNG injection rate was increased. The slope of the curves was steeper when the injection timing was delayed. For a given increase in the CNG injection rate, the increase in delay in ignition timing was higher when the degree of stratification was increased. That is both the injection rate and the degree of stratification of CNG had significant effects on the ignition timing when operated with φg = 0.20. However, the maximum total equivalence ratio was less than that obtained with CNG injection at 300 and 240° BTDC.

It was found that the combustion duration was reduced when CNG injection rate was increased at 300, 240, and 80° BTDC. When CNG was injected at 180 and 120° BTDC, the combustion duration was marginally affected and it initially decreased up to certain values of CNG injection rate and then it increased again.

Figures 15–18 show the rate of heat release and pressure rise at various injection timings. Increasing the rate of CNG injection at 300° BTDC was limited by knocking as shown in **Figure 14**. But with later injection timings, with φg = 0.20, any increase in CNG injection rate resulted in a delayed autoignition and reduced peak pressure. Therefore, increasing CNG injection rate beyond certain levels led to misfire or no fire, thereby defining the maximum load limit.

As shown in **Figure 15**, when CNG injection rate was increased, it resulted in delayed ignition timing. Up to φTotal = 0.33, the resultant peak pressure increased, and with a further increase in CNG injection rate, it decreased. Also, above φ Total = 0.33, the delay in ignition timing was more significant and resulted in decreased peak pressures. As will be discussed later in this section, combustion efficiency of both the fuels increased and CH4 emissions decreased with an increase in φTotal above 0.33 as shown in **Figures 23** and **29**.

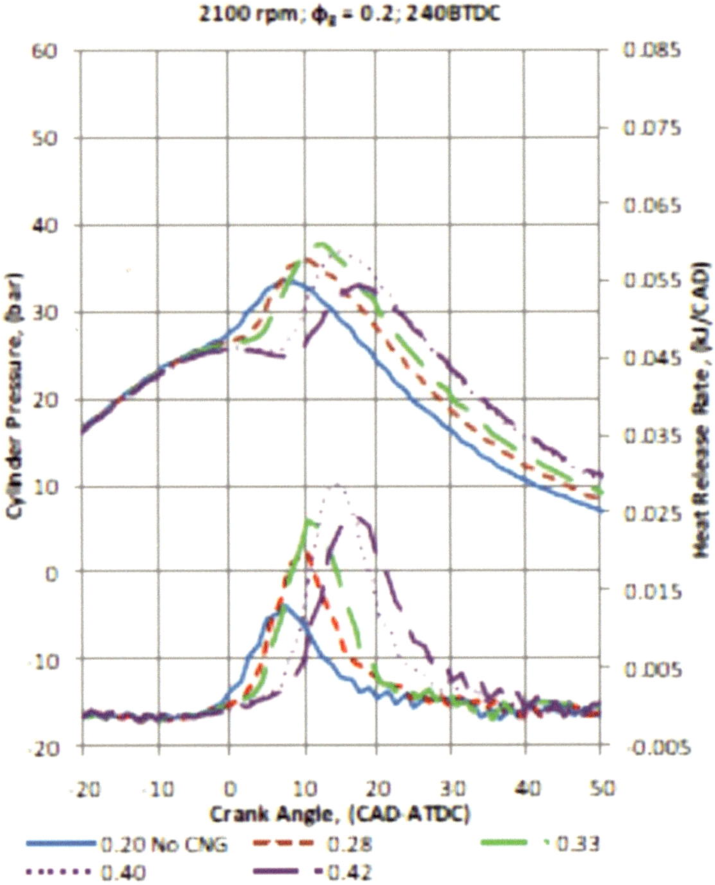

Figure 15. Pressure history and heat release rates with CNG injection at 240° BTDC.

Therefore, it can be concluded that, above φTotal = 0.33, the peak pressure was reduced due to delayed ignition, and the combustion was more complete with increase in injection rate at 240° BTDC. That is, increasing φg above 0.33 resulted in reduced peak pressures without a decrease in thermal efficiency as shown in **Figure 12**. Heat release rates increased with an increase in CNG injection rate up to φTotal = 0.40 above which it reduced again. Above φTotal = 0.42, increasing CNG injection rate resulted in misfire or no fire, and both gasoline and CNG combustion was quenched.

With CNG injection at 180° BTDC, increase in CNG injection rate resulted in a more significant delay in ignition timing. There was a marginal increase in peak pressure when φTotal was increased to 0.26 above which it reduced again. Thermal efficiency and combustion efficiency increased primarily due to the remarkable increase in the completeness of CNG combustion as suggested by CH4 emissions as shown in **Figure 29**. Heat release rate increased with an increase in CNG injection rate as shown in **Figure 17**. However, increasing CNG injection rate above φTotal = 0.26 resulted in decreased overall combustion efficiency and high CH4

http://dx.doi.org/10.5772/intechopen.72880

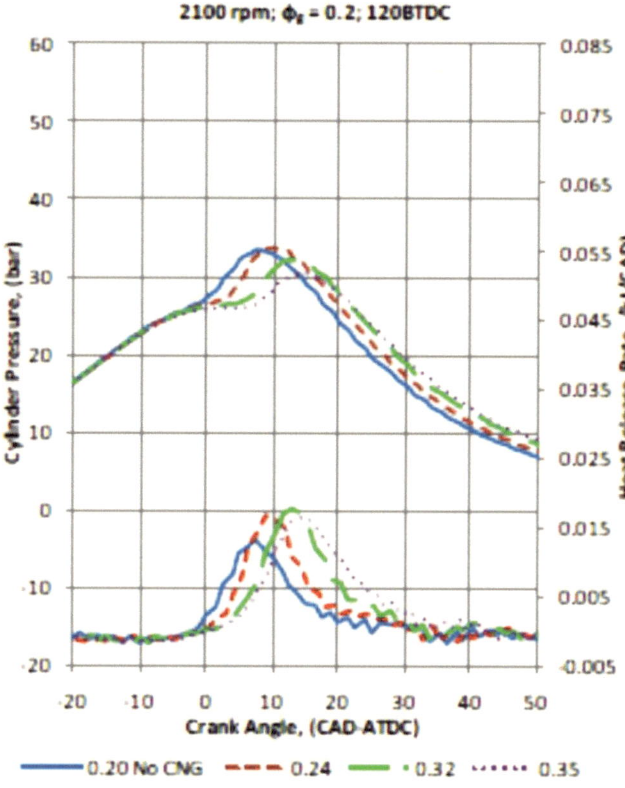

Figure 16. Pressure history and heat release rates with CNG injection at 120° BTDC.

emissions as shown in **Figures 23** and **29**. This suggests that the degree of stratification created at 180° BTDC injection results in deterioration in combustion and leads to decrease in thermal efficiency as shown in **Figure 12**. Similar, trends were observed with CNG injection at 120° BTDC when φTotal was increased above 0.24 as shown in **Figure 16**.

When CNG injection was retarded to 80° BTDC, increase in injection rate resulted in significant delay in ignition; however, there was less noticeable effect on peak pressures up to φTotal = 0.28. Increasing φTotal above resulted in a more significant delay in ignition and peak pressure, and heat release rates increased. Thermal efficiency and combustion efficiency increased primarily due to remarkable increase in completeness of CNG combustion as suggested by CH4 emissions as shown in **Figure 29**.

As shown in **Figure 19**, increasing CNG injection rate at 240° BTDC resulted in delayed ignition. At φTotal = 0.28 and 0.33, there was a slight increase in burning rate at the last stage of combustion compared to combustion with pure gasoline. At 180 and 120° BTDC, there was no significant effect on the burning rate of the fuels due to increase in CNG injection rate, but it caused a significant delay in ignition as shown in **Figures 20** and **21**. Similar results were obtained with CNG injection at 80° BTDC; however, at φTotal = 0.28 and 0.33, the combustion was slower at the initial stages and was faster at latter stages as shown in **Figure 22**.

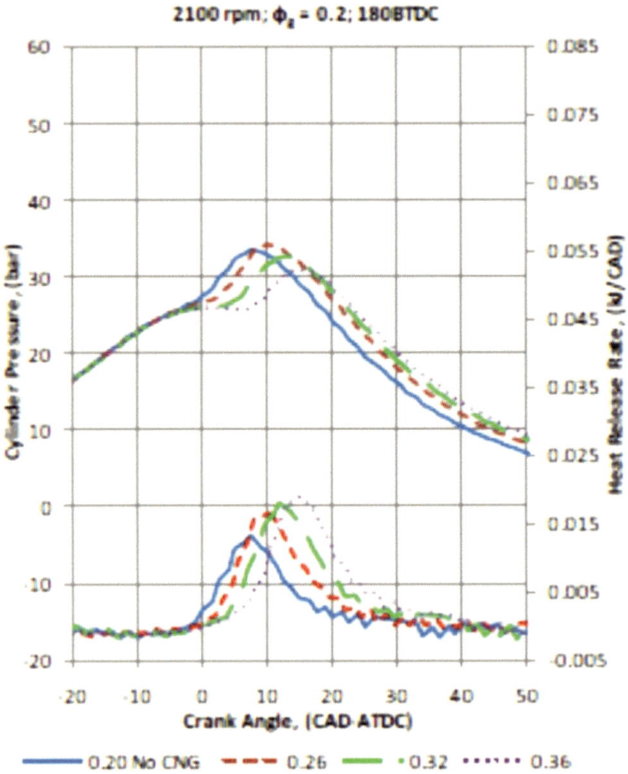

Figure 17. Pressure history and heat release rates with CNG injection at 180° BTDC.

As shown in **Figure 23**, with an increase in φTotal by CNG injection at 300, 240, and 80° BTDC, combustion efficiency increased. The highest increment was obtained with CNG injection at 80° BTDC for a given increase in φTotal due to mixture stratification. However, CNG injection at 180 and 120° BTDC, combustion efficiency increased initially but decreased again and was below 80% for all φTotal.

The exhaust gas temperature was observed to increase as the CNG injection rate was increased as shown in **Figure 24**. The increase in exhaust gas temperature with increasing CNG injection rates at 180 and 120° BTDC was less than that observed with increasing CNG injection rates at 300, 240, and 80° BTDC. When the fuels were homogeneously mixed, it resulted in higher exhaust gas temperatures due to rapid burning. Similarly, when CNG was highly stratified, it also led to higher exhaust gas temperatures.

Figure 25 shows the indicated specific NO_x ($ISNO_x$) emissions. The NO_x emissions were marginally affected and were around the same levels for all test conditions. However, different trends were observed at different injection timings and CNG injection rates.

Increasing the CNG injection rate resulted in a drastic increase in the NO_2/NO_x ratio up to a certain point and then it decreased. As shown in **Figure 26**, the ratio of NO_2/NO_x almost doubled when the CNG injection rate was increased to around φTotal = 0.33 before decreasing again. That is, up to a certain value of CNG injection rate, CNG reduced the combustion temperature and led to formation of higher amounts of NO_2.

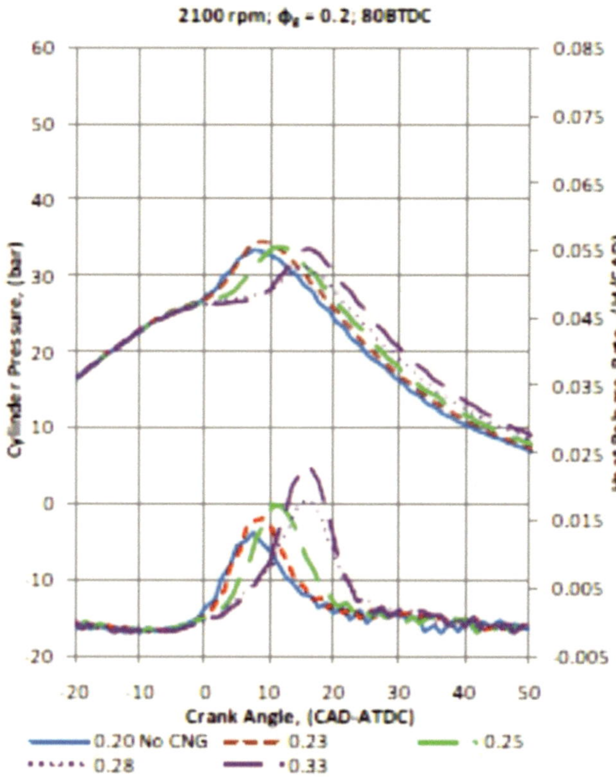

Figure 18. Pressure history and heat release rates with CNG injection at 80° BTDC.

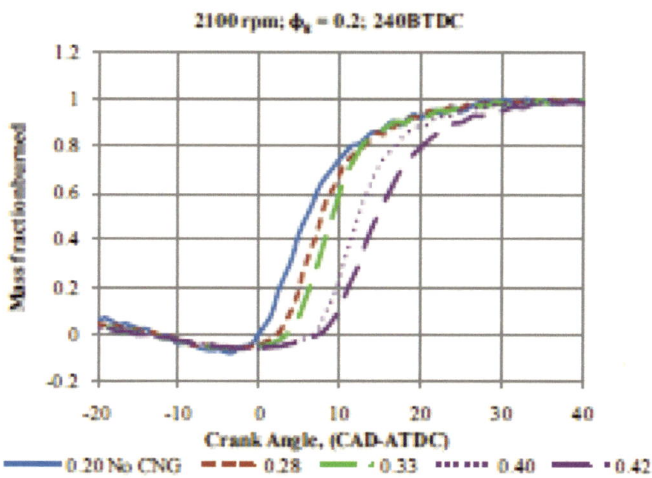

Figure 19. Mass fractions burned with CNG injection at 240° BTDC.

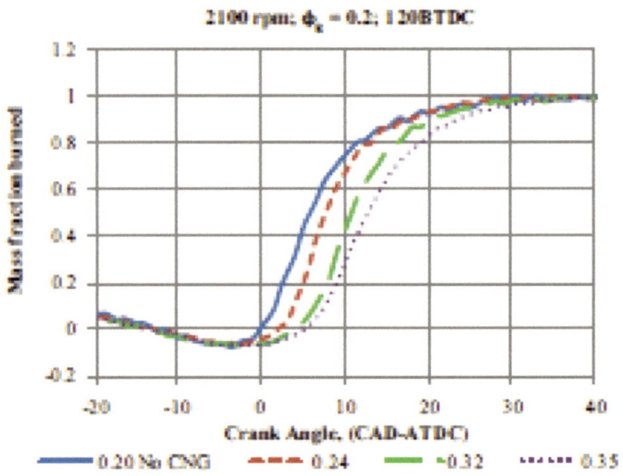

Figure 20. Mass fractions burned with CNG injection at 180° BTDC.

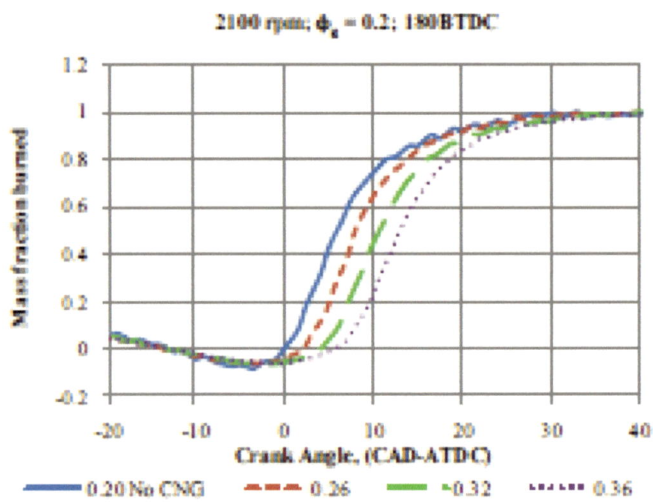

Figure 21. Mass fractions burned with CNG injection at 120° BTDC.

The indicated specific CO (ISCO) emissions were reduced significantly as the mixture was enriched with CNG by direct injection at all injection timings as shown in **Figure 27**. However, the reduction obtained was the highest when CNG was injected at 300 and 240° BTDC. Any increase in CNG injection rate at later injection timings resulted in less reductions in CO emissions. The lowest reduction was obtained at the injection timing of 80° BTDC, as the high degree of stratification of CNG limited the availability and distribution of oxygen and temperature differences in the CNG and air particles.

The HC emissions were found to be significantly affected by the degree of stratification of CNG as shown in **Figure 28**. The highest reduction in HC emissions was obtained with CNG

http://dx.doi.org/10.5772/intechopen.72880

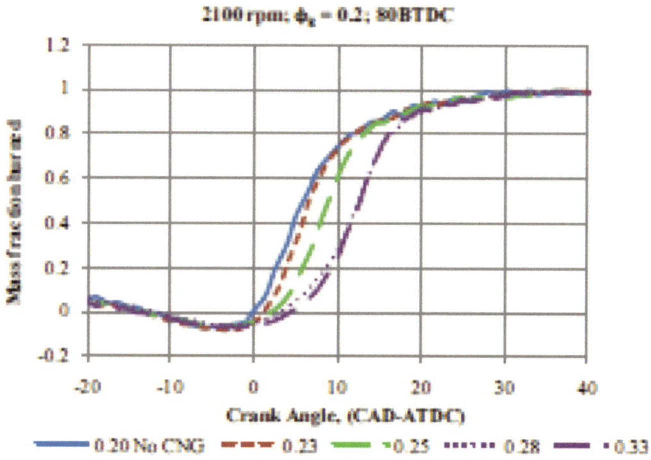

Figure 22. Mass fractions burned with CNG injection at 80° BTDC.

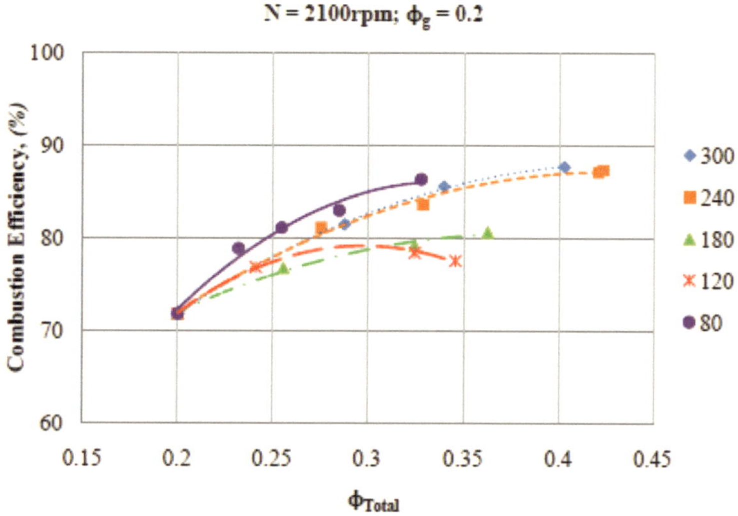

Figure 23. Effect of degree of stratification on combustion efficiency.

injection at 80° BTDC. Higher degrees of stratification of CNG resulted in more complete combustion.

Figure 29 shows the mass ratio of flow rates of CH4 in the exhaust emissions and mass flow rate of CNG injected into the cylinder. At a given constant gasoline equivalence ratio of φg = 0.20, CNG direct injection at 80° BTDC resulted in the least emission of CH4. Therefore, the combustion of CNG was more complete when it was stratified. CNG injection at 300 and 240° BTDC resulted in moderate levels of CH4 emissions, and highest values were obtained with CNG injection at 180° BTDC. This was due to the turbulence created and mixing conditions in the cylinder when the piston changed its direction at 180° BTDC.

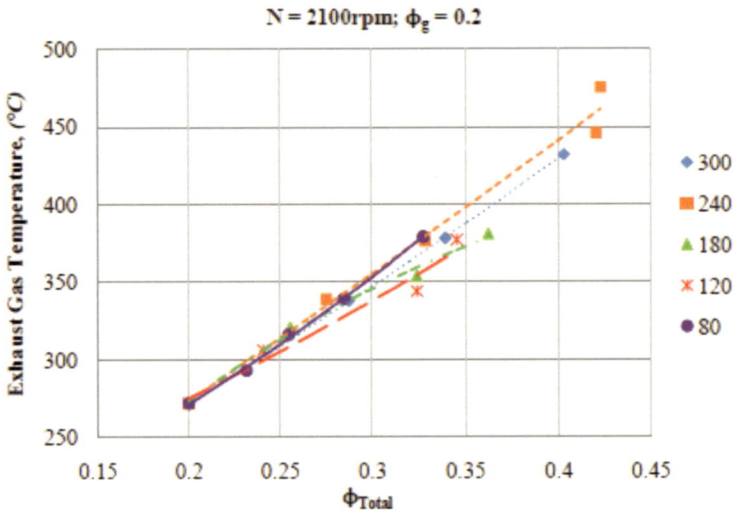

Figure 24. Effect of CNG injection on the exhaust gas temperature.

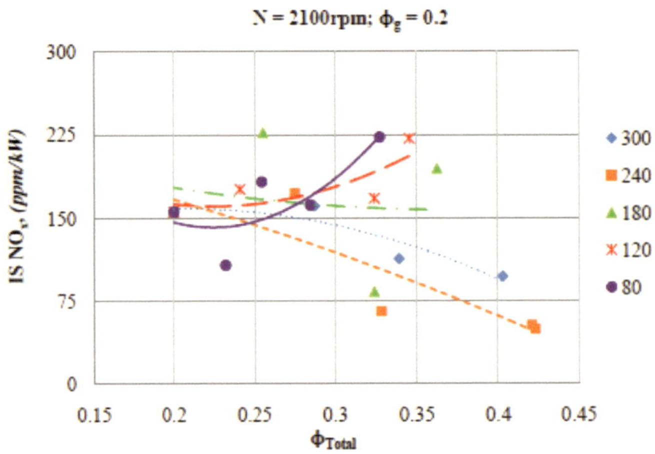

Figure 25. Effect of degree of CNG stratification on NO_x emission.

3.2. Gasoline-CNG combustion behavior in constant volume combustion chamber

The effect of injection gap on the gasoline-compressed natural gas mixture (GCNG) mixture combustion is discussed below. The injection gap alteration gave direct impact on the mixture distribution inside the combustion chamber. There are five injection gaps tested, 0, 5, 10, 15, and 20 ms. These injection gaps are expected to be able to give direct control to the mixture distribution inside the chamber.

The effect of injection gaps is shown in **Figure 30**. It shows two mixture compositions, 50 and 90% GCNG composition. The injection gap gives different effect between the two compositions. In 50% GCNG, longer injection gap gives higher combustion efficiency, maximum pressure, total heat released (THR) and shorter delay. Furthermore, it also shows longer duration for

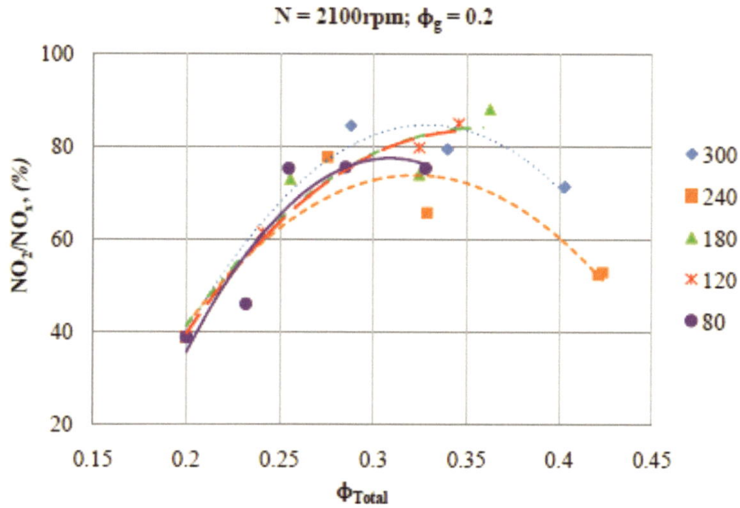

Figure 26. Effect of degree of CNG stratification on NO_2 formation.

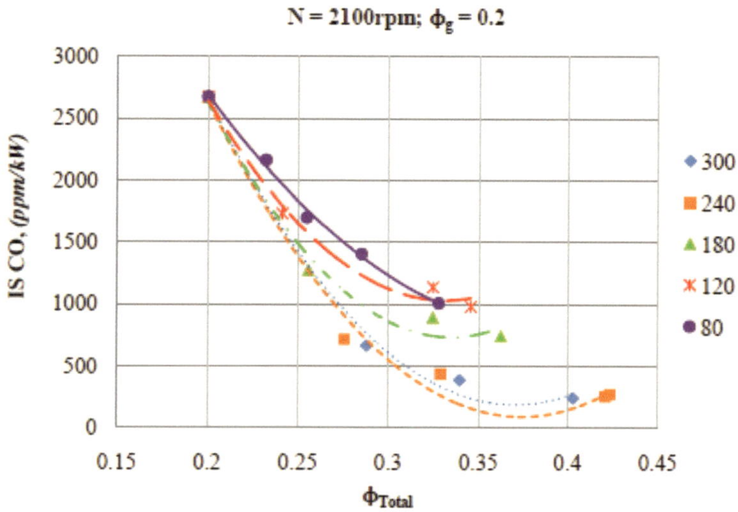

Figure 27. Effect of degree of CNG stratification on CO emissions.

all combustion stages. While on the contrary, longer injection gap reduces the combustion efficiency, maximum pressure, THR and longer combustion delay for 90% GCNG. However, the trends of the combustion duration are similar, longer duration for longer injection gap. **Figure 31** confirmed the variation of injection gap effect to the combustion process of the GCNG mixture. The turning point is shown between 70 and 80% GCNG mixture compositions. For all the mixture above 80% shows decreasing combustion efficiency with the increase in injection gaps which is contrary with the mixtures below 70% that show an increment of combustion efficiency with the increase of injection gaps. These differences may cause by the mixture distribution inside the chamber.

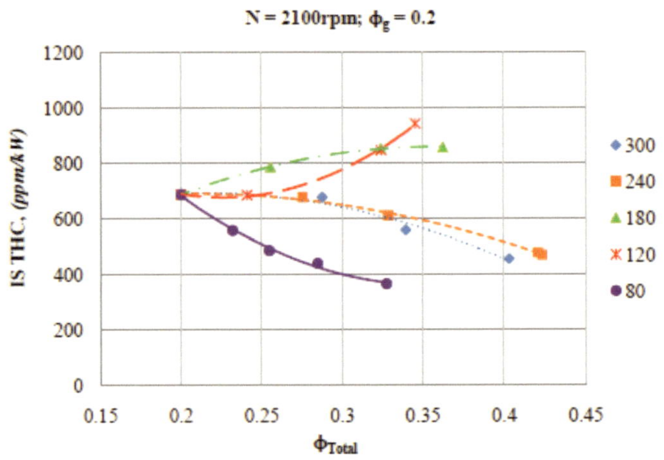

Figure 28. Effect of degree of CNG stratification on HC emissions.

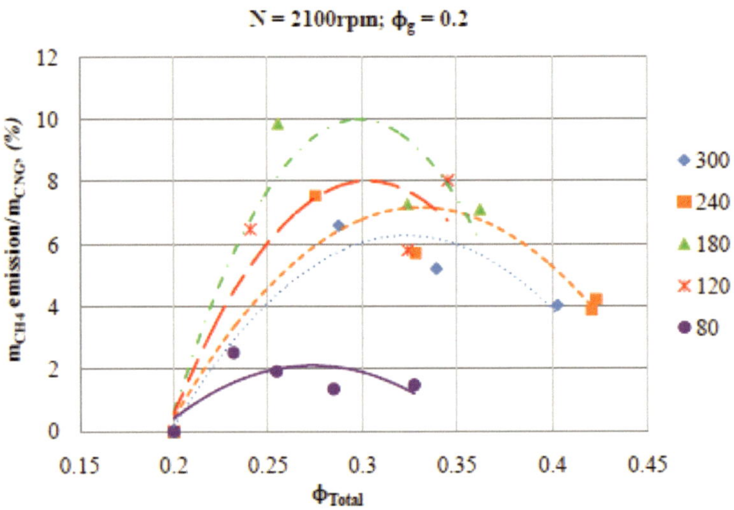

Figure 29. Effect of injection timing on the emission of CH4 with φg = 0.20.

The mixture distribution inside the chamber for GCNG mixture 30 and 90% with 0 and 20 ms injection gaps are depicted in **Figure 32**. In the figure, highly stratified mixture for 30% GCNG mixture with 0 ms injection gap. The stratification is marked by darker color on the bottom of the chamber that indicates high density fluid (gasoline). The image shows that most of the gasoline was collected at the bottom of the chamber because of the momentum of CNG injection that prevents the gasoline from reaching the top side of the chamber. 20 ms injection gap, on the other hand, shows better fuel mixing shown by fairly similar image intensity throughout the chamber.

The injection gaps for 30% GCNG mixture improve the mixing rate thus increase the combustion performance. Furthermore, the gasoline fuel is mainly accumulating at the bottom side

http://dx.doi.org/10.5772/intechopen.72880

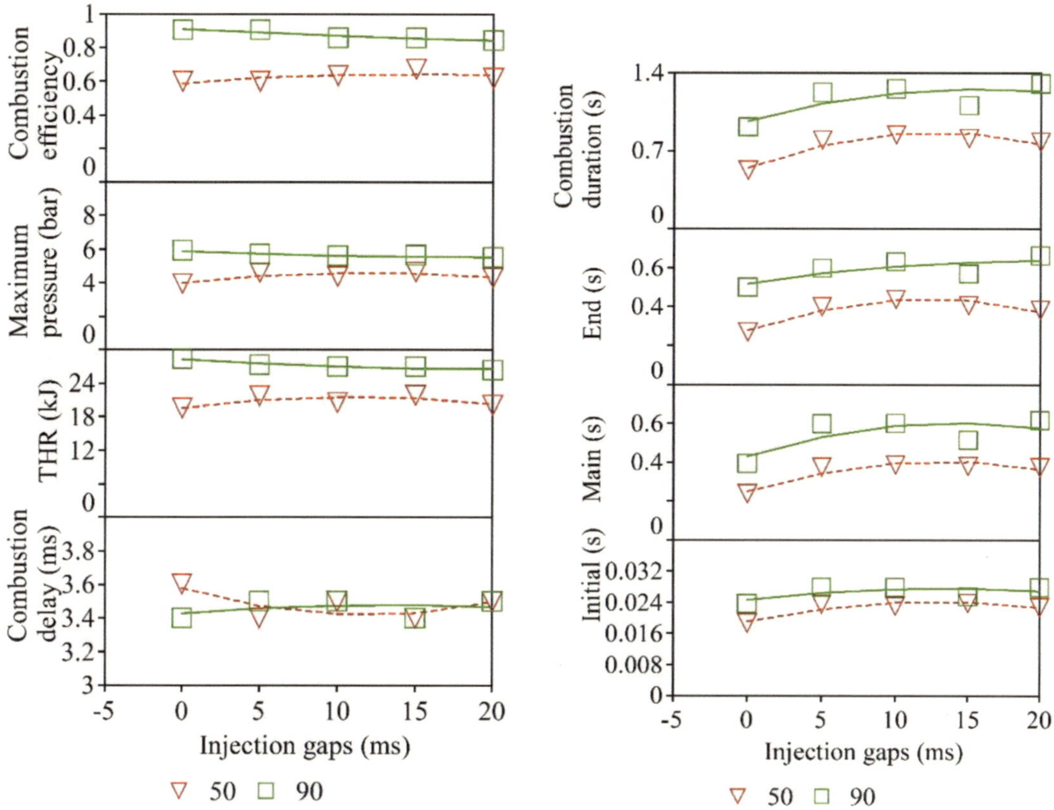

Figure 30. Effect of injection gaps on the combustion characteristics of GCNG mixture for 50 and 90% composition at lambda 1.

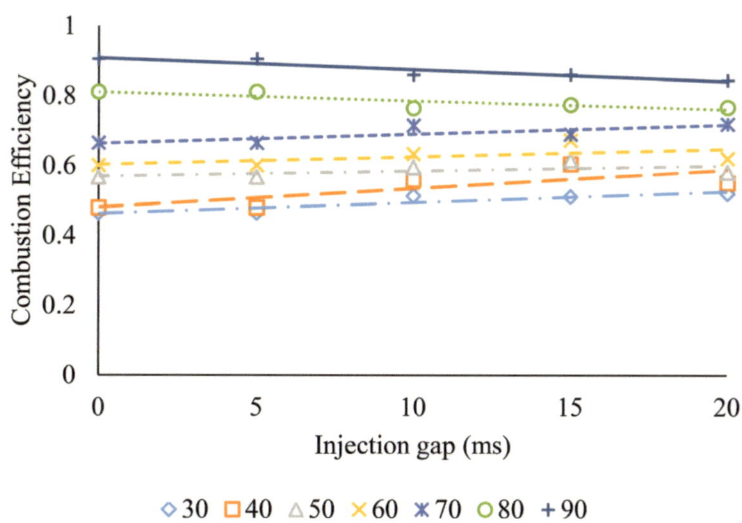

Figure 31. Effect of injection gap to the combustion efficiency for various mixture compositions at lambda 1.

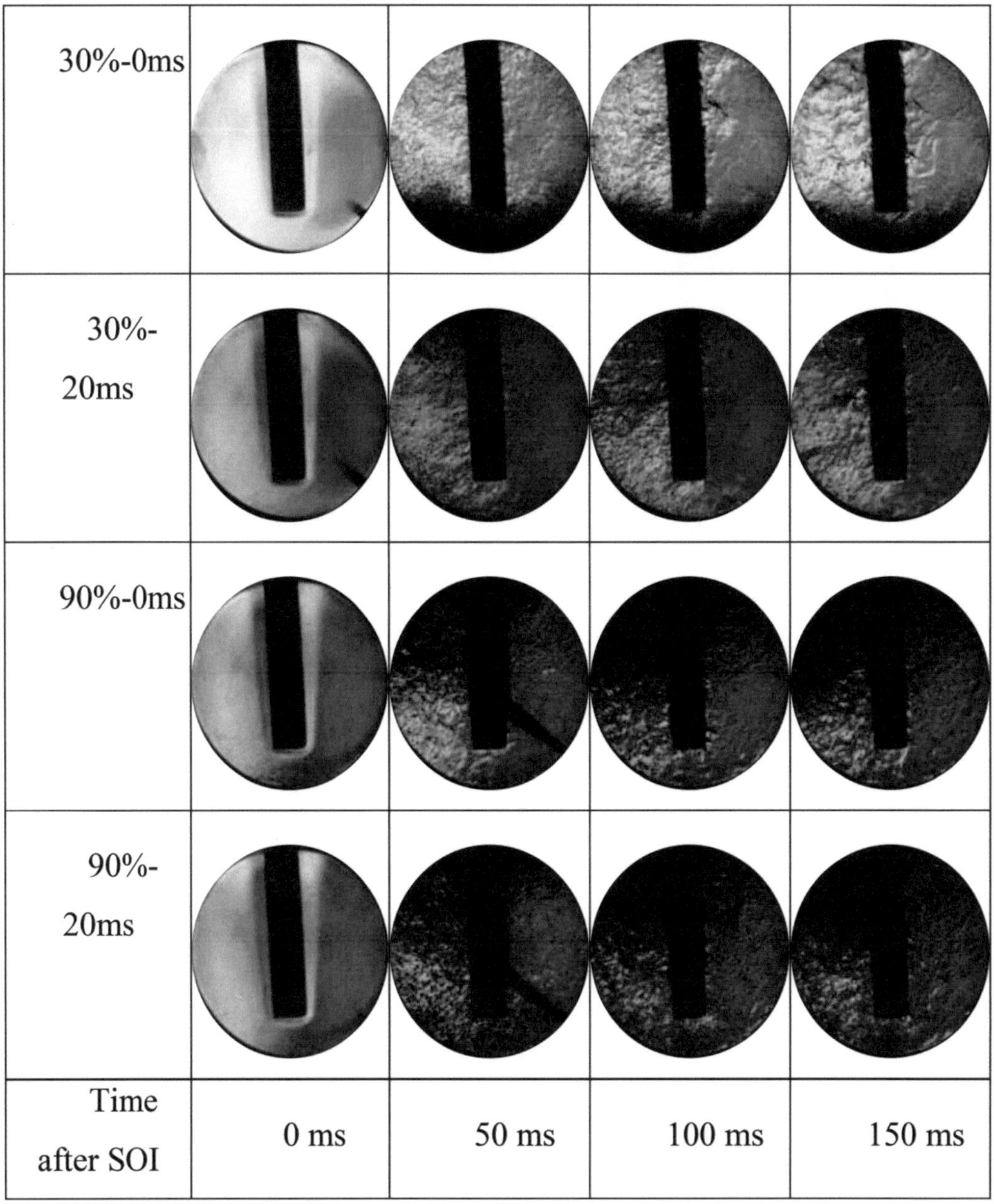

Figure 32. Mixture distribution for 30 and 90% mixture composition at 0 and 20 ms injection gaps.

which also has average low temperature compared to the top one. It makes the vaporization rate of the gasoline took longer time which also elongates the combustion delay as shown in **Figure 33**.

The injection gaps at 90% GCNG mixture, on the other hand, have similar liquid fuel distribution as in **Figure 32** where both injection gaps show concentrated fuel distribution at the top of the chamber. In spite of the similarity, the 0 ms injection gap shows the higher intensity of the liquid fuel (darker region) on the top side of the chamber compared to 20 ms injection gap. It

http://dx.doi.org/10.5772/intechopen.72880

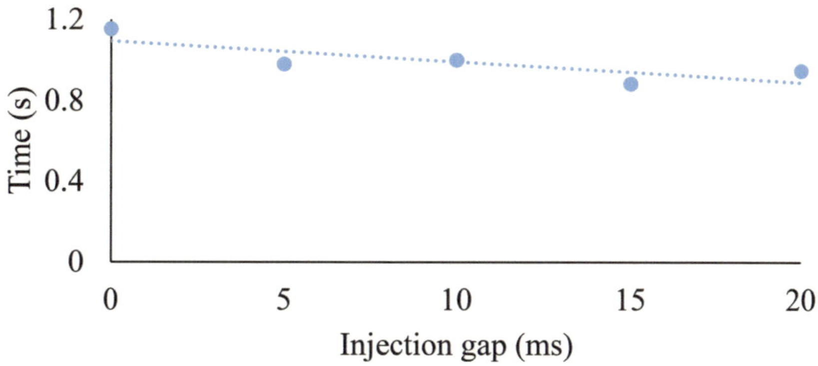

Figure 33. Combustion delay relative to SOI.

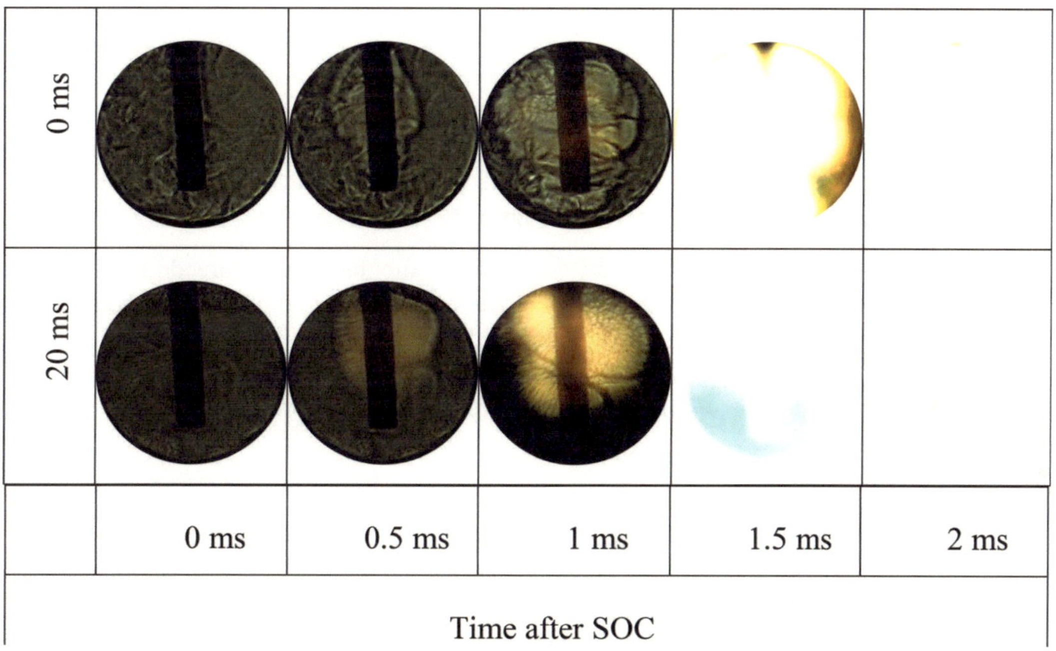

Figure 34. Combustion sequence for GCNG at 60/40 mixture composition and lambda 1.

shows that 0 ms injection gap have higher gasoline population compared to 20 ms injection gap thus the longer time required for vaporization process. This is the main reason for the lower combustion output as well as longer combustion delay for 0 ms compared to 20 ms injection gap.

The combustion sequence for GCNG at 0 and 20 ms injection gaps is depicted in **Figure 34**. The flame speed of 20 ms injection gap is faster than 0 ms injection gap with 37.02 m/s at the first 0.5 ms and 15.9 at the first 1 ms after start of combustion (SOC), while 0 ms injection gap with 30.56 m/s and 16.9 m/s at 0.5 ms and 1 ms, respectively. **Figure 34** also reveals the difference in the flame color for the two injection gaps. A 20 ms injection gap shows light blue color yet with

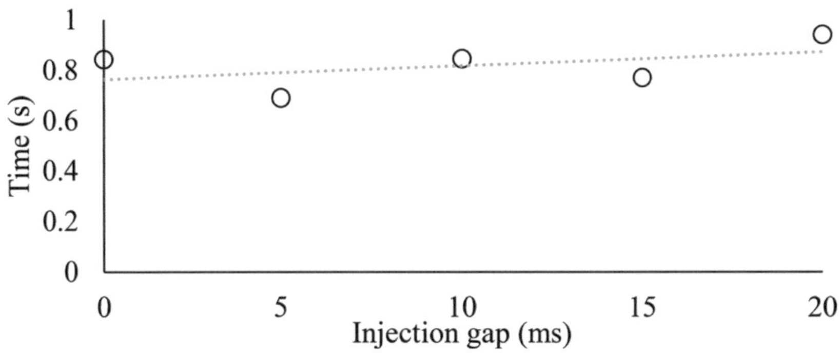

Figure 35. Effect of injection gap for 60/40 gasoline/CNG composition at lambda 1.

lower intensity compared to 100% gasoline, while 0 ms injection gap shows yellow color. It can be assumed that the blue color is the product from the same reaction that generates hydroxyl peroxide and increases the combustion output of the mixture.

In 100% gasoline combustion, the blue flame occurs because of the homogeneous mixture that creates multipoint combustion behind the flame front that significantly increases the combustion output. A similar process occurs in the 20 ms injection gaps, mixture homogeneity is achieved with this mixture as the effect of CNG injection shown by longer combustion delay relative to the start of injection as depicted in **Figure 35**.

Injection gaps proved to have a direct influence on the fuel distribution inside the chamber, thus affecting the combustion characteristics of the mixture. The combustion process in the CVC is mostly affected by the characteristics of the fuel distribution inside the chamber at the time the combustion occurs. The injection gaps, in this case, highly affect the mixture distribution inside the chamber where longer gap promotes the mixing and create a mixture that is more homogeneous.

4. Conclusions

1. It was found that more amount of gasoline was needed to ignite stratified CNG. Increasing the degree of CNG stratification resulted in delayed ignition and reduced peak pressures. The combustion efficiency of overall mixture and that of CNG were higher at high degree of CNG stratification. When the degree of CNG stratification was increased, significant reduction in HC emissions could be achieved even at low gasoline equivalence ratios. However, NO_x and CO emissions increased especially at low engine loads and high speeds. Maximum load attainable at low engine speeds could be extended by increasing the degree of CNG stratification.

2. Injection gap capable of controlling the stratification level in the CVC shown by longer injection gap produces relatively homogeneous mixture compared to the short injection gaps. The results show that the fuel stratification level has a less significant effect on the combustion stages although it has a significant effect on the combustion phasing. Stratified mixture produces shorter combustion duration especially on the main and final combustion

stages, while homogeneous mixture produces longer duration. This trend is applicable for both GCNG and DCNG mixtures.

Author details

Firmansyah[1], Abdul Rashid Abdul Aziz[2]*, Morgan Raymond Heikal[1,3], Ezrann Zharif Zainal Abidin[1] and Naveenchandran Panchatcharam[4]

*Address all correspondence to: rashid@utp.edu.my

1 Centre for Automotive Research and Electric Mobility, Universiti Teknologi Petronas, Malaysia

2 Institute of Transport and Infrastructure, Universiti Teknologi Petronas, Malaysia

3 Advanced Engineering Centre, University of Brighton, UK

4 Bharath University, India

References

[1] Aoyama T, Hattori Y, Mizuta J. An experimental study on premixed-charge compression ignition gasoline engine. JSAE Review. 1995;**16**(3):314

[2] Kokjohn SL, Hanson RM, Splitter D a, Reitz RD. Fuel reactivity controlled compression ignition (RCCI): A pathway to controlled high-efficiency clean combustion. International Journal of Engine Research. 2011;**12**(3):209-226

[3] Musculus M. Combustion Regime Visualization. 2011. [Online]. Available: http://www.sandia.gov/ecn/tutorials/visualization.php

[4] Stanglmaier RH, Roberts CE. Homogeneous charge compression ignition (HCCI): Benefits, compromises, and future engine applications. SAE Technical Paper Series. vol. 1999-01-36; 1999. pp. 1999-01-3682

[5] Saitou K, Iijima A, Otagiri Y, Yoshida K, et al. A study of ignition characteristics of an HCCI engine operating on a two-component fuel. SAE. 2010

[6] Sjöberg M, Lars-Olof E, Eliassen T, Magnusson L, Ãngstrom HE. GDI HCCI: Effects of injection timing and air swirl on fuel stratification, combustion and emission formation. SAE Technical Paper Series. vol. 2002-01-01; 2002

[7] Agarwal AK, Singh AP, Maurya RK. Evolution, challenges and path forward for low temperature combustion engines. Progress in Energy and Combustion Science. 2017;**61**:1-56

[8] Wang Y, Yao M, Li T, Zhang W, Zheng Z. A parametric study for enabling reactivity controlled compression ignition (RCCI) operation in diesel engines at various engine loads. Applied Energy. 2016;**175**:389-402

[9] Wu Y, Hanson R, Reitz RD. Investigation of combustion phasing control strategy during reactivity controlled compression ignition (RCCI) multicylinder engine load transitions. Journal of Engineering for Gas Turbines and Power. 2014;**136**(9):91511

[10] Fang W, Kittelson DB, Northrop WF. Optimization of reactivity-controlled compression ignition combustion fueled with diesel and hydrous ethanol using response surface methodology. Fuel. Jul. 2015

[11] Li J, Yang W, Zhou D. Review on the management of RCCI engines. Renewable and Sustainable Energy Reviews. 2017;**69**(May 2016):65-79

[12] Zhou DZ, Yang WM, An H, Li J, Shu C. A numerical study on RCCI engine fueled by biodiesel/methanol. Energy Conversion and Management. 2015;**89**:798-807

[13] Li Y, Jia M, Chang Y, Liu Y, Xie M, Wang T, Zhou L. Parametric study and optimization of a RCCI (reactivity controlled compression ignition) engine fueled with methanol and diesel. Energy. 2014;**65**:319-332

[14] Indrajuana A, Bekdemir C, Luo X, Willems F. Robust multivariable feedback control of natural gas-diesel RCCI combustion. IFAC-Papers OnLine. 2016;**49**(11):217-222

[15] Nazemi M, Shahbakhti M. Modeling and analysis of fuel injection parameters for combustion and performance of an RCCI engine. Applied Energy. 2016;**165**:135-150

[16] Benajes J, Pastor JV, García A, Monsalve-Serrano J. The potential of RCCI concept to meet EURO VI NOx limitation and ultra-low soot emissions in a heavy-duty engine over the whole engine map. Fuel. 2015;**159**:952-961

[17] Paykani A, Kakaee A-H, Rahnama P, Reitz RD. Effects of diesel injection strategy on natural gas/diesel reactivity controlled compression ignition combustion. Energy. 2015; **90**(Part 1):814-826

[18] Kokjohn SL, Musculus MPB, Reitz RD. Evaluating temperature and fuel stratification for heat-release rate control in a reactivity-controlled compression-ignition engine using optical diagnostics and chemical kinetics modeling. Combustion and Flame. Jun. 2015; **162**(6):2729-2742

[19] Pedrozo VB, May I, Lanzanova TDM, Zhao H. Potential of internal EGR and throttled operation for low load extension of ethanol–diesel dual-fuel reactivity controlled compression ignition combustion on a heavy-duty engine. Fuel. 2016;**179**:391-405

[20] Kakaee A-H, Nasiri-Toosi A, Partovi B, Paykani A. Effects of piston bowl geometry on combustion and emissions characteristics of a natural gas/diesel RCCI engine. Applied Thermal Engineering. 2016;**102**:1462-1472

[21] Benajes J, Pastor JV, García A, Monsalve-Serrano J. An experimental investigation on the influence of piston bowl geometry on RCCI performance and emissions in a heavy-duty engine. Energy Conversion and Management. Oct. 2015;**103**:1019-1030

[22] Genchi G, Pipitone E. Preliminary experimental study of double fuel HCCI combustion. Energy Procedia. 2015;**81**:784-793

Chapter 5

Remote Combustion Sensing in Diesel Engine via Vibration Measurements

Ornella Chiavola, Erasmo Recco and Giancarlo Chiatti

Additional information is available at the end of the chapter

http://dx.doi.org/10.5772/intechopen.69761

Abstract

An efficient control of the combustion process is required in order to comply with regulations on pollutant emissions from internal combustion engines. Literature presents investigations devoted to explore the potentiality of externally mounted sensor (speed sensor, microphone, and accelerometer) for combustion diagnosis. A relationship exists between the combustion event measured via an in-cylinder pressure transducer and engine block vibration measured via an accelerometer. Time and frequency domain processing of acquired signals highlighted the correlation between parameters able to characterize the combustion development and features derived from the engine block vibration data. A methodology was developed by the authors that demonstrated to be suitable for real-time estimation of combustion progress based on engine vibration. A two-cylinder common rail diesel engine of small displacement was tested; two configurations were investigated, naturally aspirated, and turbocharged. The in-cylinder pressure and block vibration signals were acquired and processed in time and frequency domains. The vibrational components mainly related to the combustion process were extracted, and indicators of the combustion positioning were computed. The angular positions of start of combustion (SOC) and MFB50 computed via the heat release curve by means of the in-cylinder pressure measurements were compared to those obtained by means of the accelerometer signal. High correlation coefficients were obtained for the data acquired during the testing of both naturally aspirated and turbocharged configurations in the complete engine operative field.

Keywords: diesel engine, combustion, in-cylinder pressure, engine vibration, non-intrusive measurements

1. Introduction

Future regulations on internal combustion engine will require continuous tightening of pollutant emissions from internal combustion engines. Literature highlights considerable research activity on combustion monitoring and closed-loop control systems in order to ensure improvement of exhaust and noise emissions and reduction of fuel consumption.

Due to the strong dependence of combustion characteristics (ignition delay, pressure rise rate, peak pressure, and combustion duration) from injection settings, algorithms for closed-loop combustion control via injection parameters have been developed [1–3]. In these algorithms, the target values are mapped versus load, speed, and other factors in order to optimize emissions/performance despite changes in fuel properties and engine aging.

Even if low-cost transducers for in-cylinder pressure measurements have been recently proposed, limitations related to reliability, lifetime caused by the harsh environment, and mounting problems still represent limiting factors for their employment.

Advanced methodologies have been proposed in which nonintrusive measurements are used to evaluate quantities able to provide information about combustion progress. These measurements offer the advantages of guaranteeing the absence of any type of interaction with the engine operation. Moreover, the sensors can be installed in any type of engine without the need of modification.

Among them, the most promising approaches are based on engine speed fluctuation, noise radiation, and vibration measurements. Crankshaft angular speed, noise emission, and vibration measurement methods have been proposed to be used during development and calibration stages of the engine and for onboard application to control the combustion progress.

The fluctuating waveform of crankshaft angular speed versus angle is caused by the imbalance between combustion torque and external torque. Engine speed frequency component signal has demonstrated to be correlated with the torque signal. Several methods have been proposed to extract from the crankshaft angular speed measurement information about the combustion progress. Ponti et al. [4] estimated the position where 50% of mass is burnt inside the cylinder starting from the instantaneous engine speed fluctuation analysis. Moro et al. [5] presented a method for in-cylinder pressure reconstruction based on engine speed signals. Taglialatela et al. [6] proposed a model to estimate the combustion quality by means of the processing of crankshaft speed signals. Desbazeille et al. [7] developed a methodology for combustion diagnosis via the angular speed variations.

Methodologies have been proposed for combustion monitoring via engine noise radiation. Microphones offer the advantages to be installed at a distance from the engine. Noise emission from internal combustion engines is a very complex signal whose quality and levels are strongly reliant on the engine type and architecture. Even if the microphone signal has demonstrated to be correlated with the combustion process, it is highly contaminated by noise components caused by many overlapping sources (injection, piston slap, valves, oil pump, and turbocharger). The complex processing, required to insulate the combustion-related component from the measurements, has resulted a limited research activity on this topic. Jiang et al. [8] presented a method for diesel combustion monitoring based on acoustic measurements. Chiatti et al. [9] developed a methodology to characterize the in-cylinder pressure development by means of the engine noise emission. Gu et al. [10, 11] used acoustic measurements for condition monitoring of diesel engines. Torii [12] presented a technique to separate the engine noise radiation into the contributions of combustion and mechanical noise. Kaul et al. [13] investigated the acoustic emissions response caused by various engine cycle events.

The rapid pressure change in the cylinder during the combustion process gives rise to the engine structure vibrations. Piston slap, valves impacts, and gear transmissions are unwanted vibration sources that are responsible for components that decrease the signal-to-noise ratio. Vibration-based algorithms have been developed and proposed for indirect investigation of combustion process. Polonowski et al. [14] analyzed the signals from accelerometers positioned in multiple placements and orientations on an engine with the aim of investigating the potential of these sensors for combustion characterization. Lee et al. [15] investigated the correlation between the maximum heat release rate and the engine vibration. Jia et al. [16] proposed a neural network to correlate the engine block acceleration and the heat release rate. Jung et al. [17] performed a closed-loop control for the combustion process based on the engine vibration signals.

A methodology was developed by the authors, in which the block vibration signal from two different configurations of a two-cylinder common rail diesel engine is processed for combustion positioning within the engine cycle. The configurations were naturally aspirated and turbocharged.

These are the main steps of the methodology:

- selection of the optimal positioning for the accelerometer;
- time frequency analysis of the in-cylinder pressure and vibration signals and evaluation of their coherence function in order to select a frequency bandwidth in which spectral components of in-cylinder pressure and engine block vibration exhibit strong correlation;
- processing of the accelerometer traces to extract the vibration components mainly related to the combustion process;
- computation of indicators for combustion evolution characterization via vibration signal [start of combustion (SOC), angular position, where the 50% of mass is burnt inside the cylinder].

Results obtained in the engines complete operative ranges proved that the methodology based on vibration measurement is suitable for the real-time estimation of combustion progress.

2. Experimental setup and tests

Measurements were carried out on a two-cylinder diesel engine equipped with a common rail injection system, whose main application is in micro cars and urban vehicles (its technical data are reported in **Table 1**).

Cylinders	2
Displacement	440 cm^3
Bore	68 mm
Stroke	60.6 mm
Compression ratio	20:1
Maximum power	8.5 kW @ 4400 rpm
Maximum torque	23 Nm @ 2400 rpm

Table 1. Engine specifications.

Two configurations of the engine have been tested: naturally aspirated and turbocharged.

The engine was installed with an asynchronous motor (Siemens 1PH7, characterized by nominal torque 360 Nm and power 70 kW) in the test bed of Engineering Department at Roma Tre University. The engine was managed by a fully opened ECU, in order to control injection parameters (injection strategy, injection timing, and duration).

HBM T12 was used for torque measurement. AVL Fuel Balance 733 was used for fuel consumption measurement.

The in-cylinder pressure was measured with a piezoelectric transducer AVL GU13P (the preheating plug was substituted by the pressure probe).

The engine speed and the crank angle position were measured by the optical encoder AVL 364C. It generates transistor-transistor logic (TTL) rectangular pulse signals: one is the trigger signal that was used to compute the engine speed; the other is the code division multiplexing (CDM) signal that was set to a resolution of 0.1 crank angle degrees.

An Endevco 7240C accelerometer was used to measure the engine block vibration. It is a high-temperature piezoelectric mono-axial accelerometer with a nominal sensitivity of 3 pC/g and a resonance frequency of 90 kHz. The vibration signal was conditioned via B&K Nexus device (amplifier and low-pass filter). A preliminary investigation was devoted to select the optimal position and orientation of the accelerometer, able to guarantee high sensitivity as regards the combustion event and low sensitivity to mechanical sources. Details may be found in Ref. [18]. The accelerometer was mounted on the top of the engine block by means of a threaded pin on one of the stud that fastens the cylinder head to the block. The mounting was chosen in order to ensure a rigid connection to the structural engine members.

Figure 1 shows the engine setup and a detail of the accelerometer location.

The sampling frequency was varied according to the engine speed, thus to ensure a fixed crank angle resolution of the signals.

http://dx.doi.org/10.5772/intechopen.69761

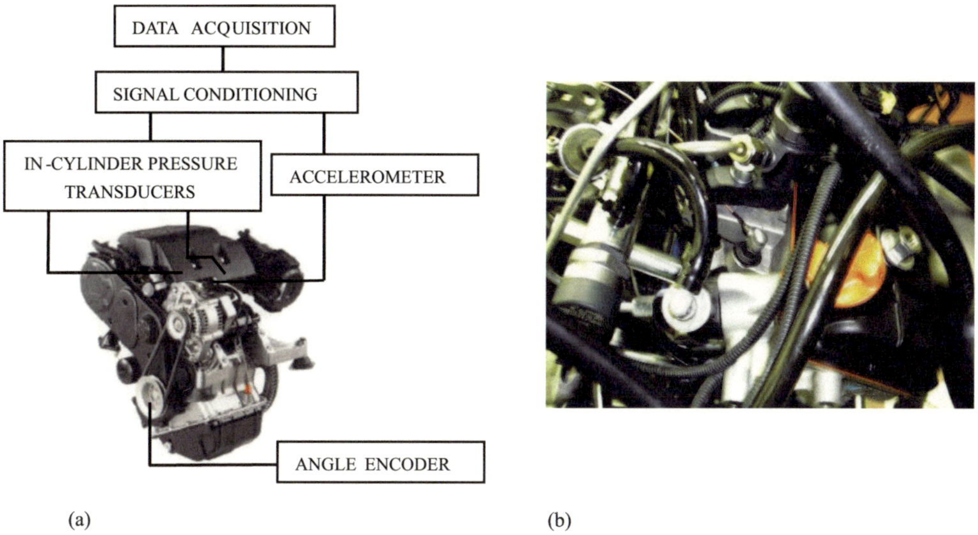

Figure 1. (a) Engine set up; (b) accelerometer location.

Data acquisition was controlled by means of LabVIEW software, by using self-developed programs. NI board types 6110, 6533, and 6259 were used.

The measurements were performed over the engine speed range 2000–4400 rpm, at different load conditions (from 50% to full load condition). For each running condition, 25 engine cycles were used to average the signal, thus to attenuate the engine cycle irregularities; the increase of such a number did not change the feature of the trends.

All acquisitions started after the engine warm-up, when the engines reached under nominally stationary conditions.

3. Results

This section focuses on the vibration and in-cylinder pressure data processing and it is devoted to describe in detail the developed methodology. In the first part, some representative crank angle evolutions of in-cylinder pressure and accelerometer signals related to naturally aspired configuration are shown and results of frequency domain analysis are presented. In the second part, results obtained with the turbocharged engine configuration are shown.

Figures 2–5 present the time-histories related to naturally aspirated configuration of the engine. The plot of **Figure 2** shows data obtained at 2000 rpm, full load condition with three different injection settings, according to **Table 2**. Case 1 was characterized by two-shot injections (pre and main injections). In cases 2 and 3, multiple injections (pilot, pre, and main injections) were imposed. These cases differentiate for the injection timings.

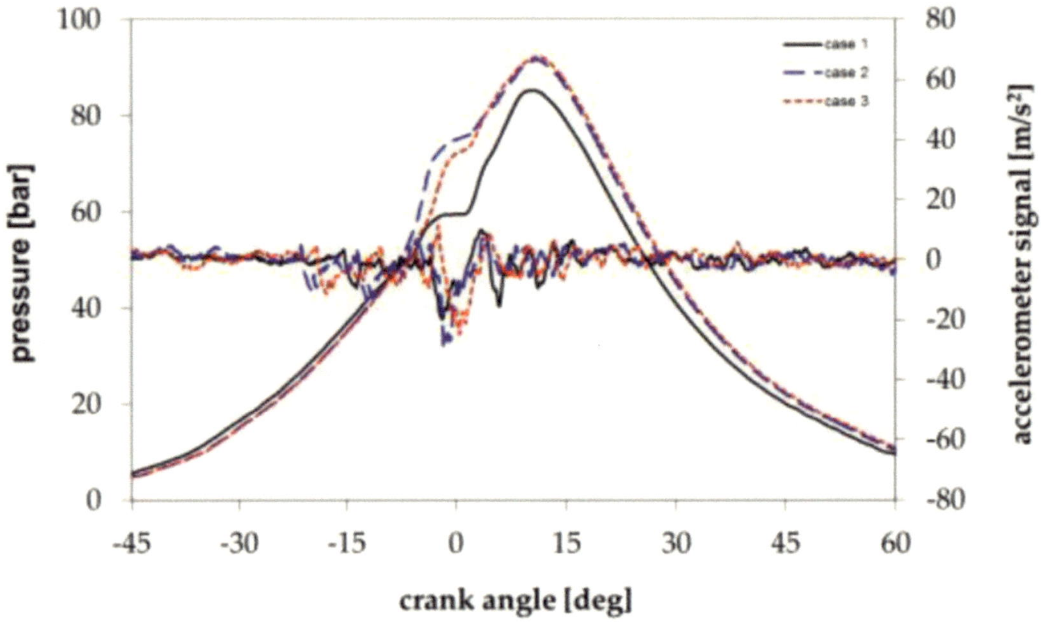

Figure 2. In-cylinder pressure and accelerometer signals at 2000 rpm, 100% load.

In-cylinder pressure traces are superimposed on the corresponding block vibration curves. The accelerometer signals highlight that the mechanical components of the engine vibration (caused by intake and exhaust valve open/close, fuel injection, and piston slap) are less evident than those related to the combustion event.

The abrupt pressure gradient due to the initial air-fuel mixture ignition is responsible for high frequency and high amplitude oscillations in the accelerometer traces, regardless of which injection setting is imposed on the engine.

As the injection parameters change, the in-cylinder pressure development modifies accordingly, and the engine vibration tunes with pressure variations for both the crank angle delay and the maxima amplitude and gradient.

Figure 3 shows the comparison between signals obtained by imposing on the engine a variation of engine speed. The main differences in the pressure development are located at the

	Q [mm³/cycle]			SOI [crank angle BthC]		
	pil	pre	main	pil	pre	main
Case 1	0	1	13.5	0	16	6
Case 2	1	1	12.5	24	16	6
Case 3	1	1	12.5	22	13	6

Table 2. Injection data at 2000 rpm, 100% load.

http://dx.doi.org/10.5772/intechopen.69761

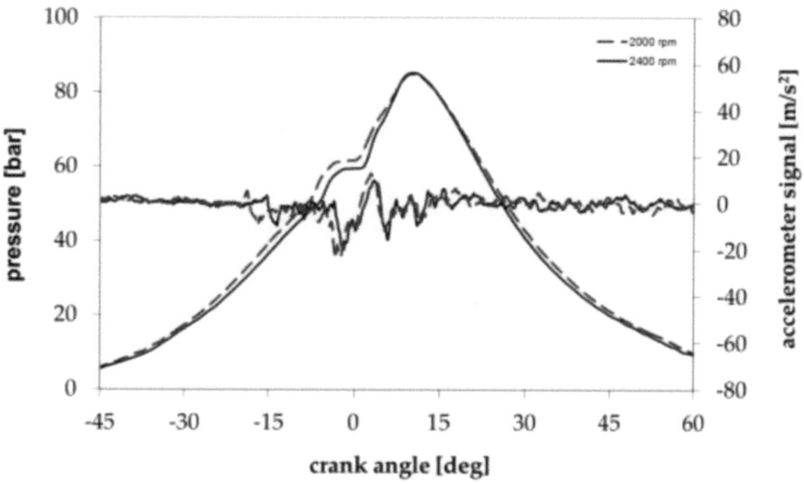

Figure 3. In-cylinder pressure and accelerometer signals at 100% load.

beginning of the combustion process and agree with the variations in the block vibration traces in the same crank angle interval. **Figure 4** shows how the pressure development and block vibration are affected by a variation of the engine load condition. The contribution of combustion process to the vibration traces is evident: accelerometer signals modify in both time and amplitude accordingly with the variations in pressure traces.

Figure 5 shows in-cylinder pressure in one cylinder and block vibration during one complete engine cycle. The plot highlights that the combustion events in both cylinders affect the accelerometer trend (the combustions have 360 crank angle degrees shift). In the crank angle intervals out of those in which combustion processes take place, a low frequency oscillation is exhibited; the frequency of this oscillation is equal to two times the engine speed value. It is to ascribe to engine mechanical components.

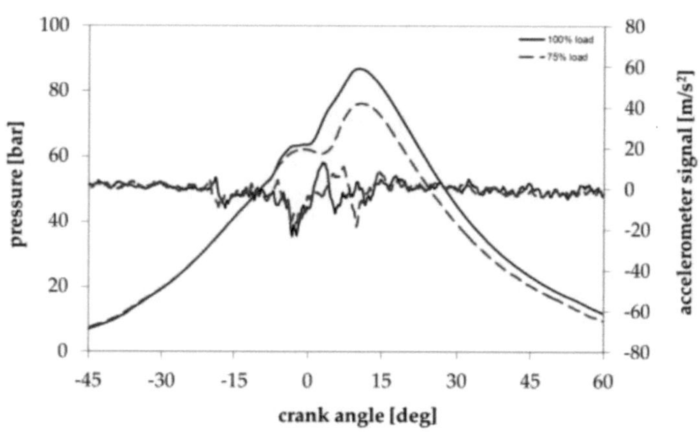

Figure 4. In-cylinder pressure and accelerometer signals at 2400 rpm.

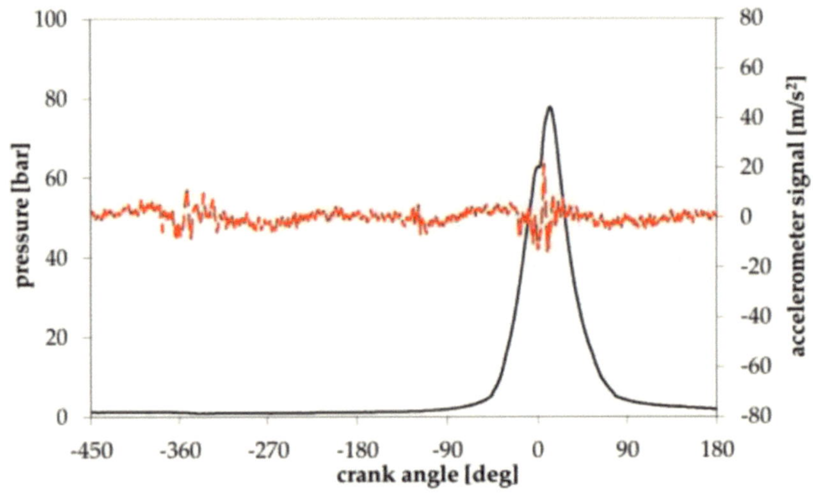

Figure 5. In-cylinder pressure and accelerometer signals at 3200 rpm, 100% load.

In order to insulate the vibration component mainly related to the combustion development, an analysis of the acquired signals in the frequency domain was performed. Coherence function between in-cylinder pressure and acceleration signals was computed. Coherence function is defined as the ratio of the cross power spectral density of an input signal (in-cylinder pressure) and the corresponding output signal (engine block vibration) to the product of the power spectral density of each signal. The function was computed by using windowed data (Hamming window 1/6 of the engine cycle long was used). Further details may be found in Ref. [19].

Figures 6 and **7** show the coherence function trends obtained at 2000 and 2400 rpm, full load condition. The plots highlight that it is possible to define a narrow frequency band

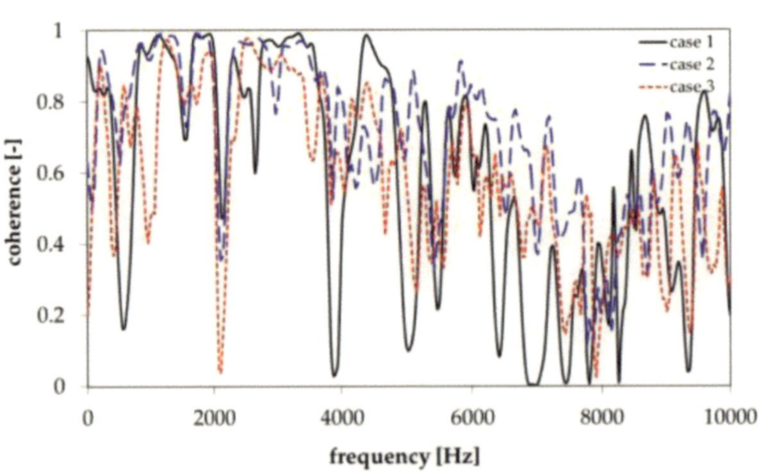

Figure 6. Coherence function at 2000 rpm, 100% load [20].

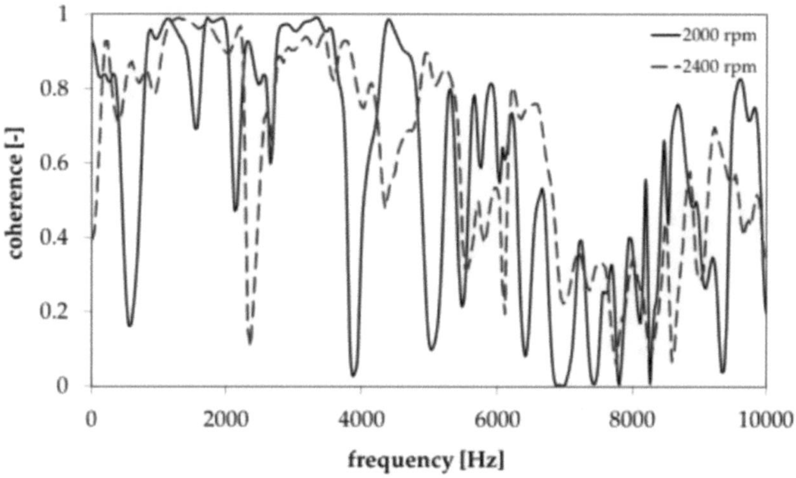

Figure 7. Coherence function at 100% load.

(approximately in the range 1000–2000 Hz) in which coherence function exhibits the highest values. Tests have been performed in order to investigate the effect of engine operative condition on the relation between in-cylinder pressures and block vibration signals. From the analysis of the coherence traces obtained in the engine complete operative field, it came out that no matter which the engine operating condition is, it is always possible to select a range of frequency values in which coherence has the highest values thus showing a linear relationship between in-cylinder pressure and block vibration signals [20, 21]. The processing of the acquired data demonstrated that load condition has a weak effect on the frequency band, whereas it is reliant on the engine speed value, in agreement with results obtained during previous experimental activity [20].

For each engine operative condition, the frequency band, in which in-cylinder pressure and accelerometer traces exhibited high values of correlations, was selected and used to band-pass filter the vibration data, thus allowing to remove from the signal all the components due to sources other than the combustion process.

Figure 8 shows the obtained filtered accelerometer signal related to 3200 rpm, full load condition. The signal is superimposed on the in-cylinder pressure trace; both trends were normalized by dividing all data for the maximum amplitude. The plot highlights oscillations of high amplitude in two-crank angle regions, corresponding to the intervals in which combustion events take place in the cylinders. These oscillations are mainly caused by the combustion since the filtration allowed to keep into the signal only the components highly correlated to the combustion.

Aimed at relating the combustion process to the filtered accelerometer trace, the rate of heat release (ROHR) was computed starting from the in-cylinder pressure, through a thermodynamic model in which the Woschni model was used for the instantaneous heat loss to the cylinder wall.

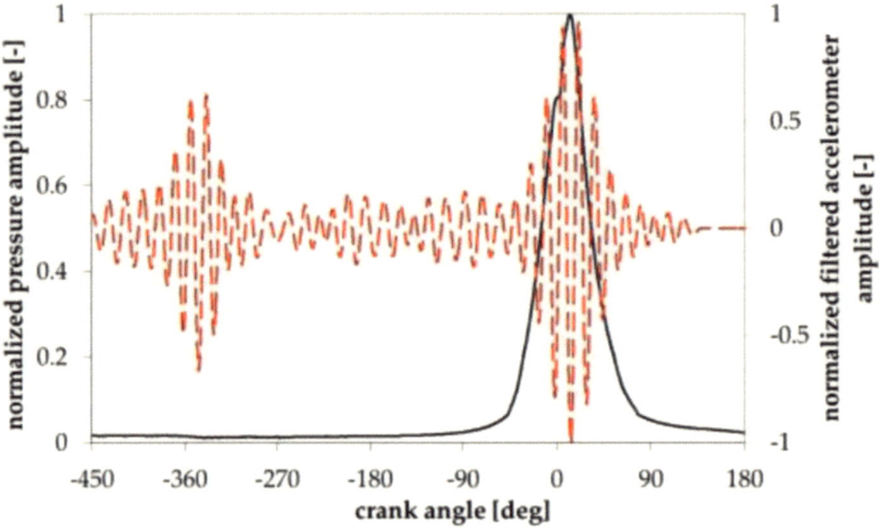

Figure 8. Normalized in-cylinder pressure and filtered accelerometer signals at 3200 rpm, 100% load.

Figure 9 shows the ROHR trace superimposed on the accelerometer signal for 3200 rpm, full load condition. Data were normalized with the maximum value. The circle in the plot highlights a zero crossing in the accelerometer trace that indicates the crank angle value corresponding to the start of combustion (SOC).

Starting from the ROHR, the cumulative heat release (CHR) was computed aimed at evaluating the crank angle values corresponding to the burnt mass fraction. **Figure 10** shows the CHR and the filtered accelerometer trend; circles are used to point out SOC and the angular position at which half of the injected fuel is burnt (MFB50).

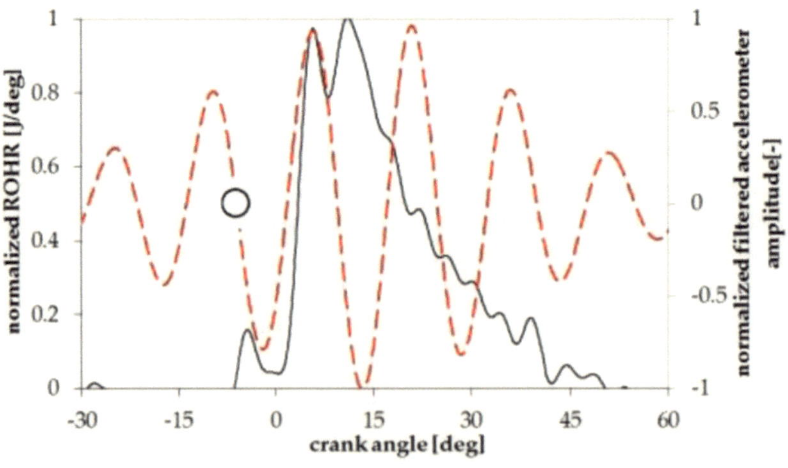

Figure 9. Normalized rate of heat release and filtered accelerometer trends at 3200 rpm, 100% load.

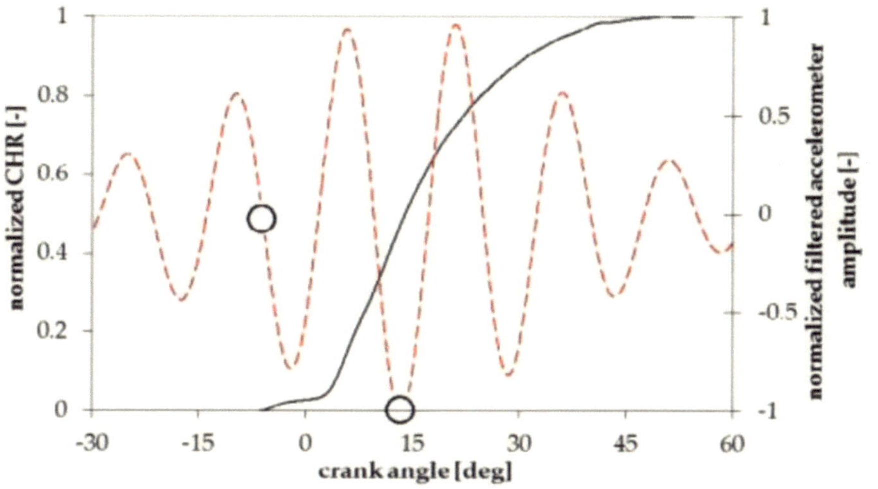

Figure 10. Normalized cumulated sum of rate of heat release and filtered accelerometer trends at 3200 rpm, 100% load.

The same processing was performed with the signals acquired with turbocharged engine configuration. **Figure 11** presents the crank angle evolution of in-cylinder pressure and accelerometer trace at 4000 rpm, 100% load.

The vibration trace appears more noisy in comparison with that one obtained during tests with naturally aspirated configuration (i.e., signals are shown in **Figures 2** and **3**), but the effect of combustion process on the accelerometer signal is evident, as shown in **Figure 12**. In the plot, the vibration signal acquired in fired condition is compared to that one related to the same engine condition but obtained with naturally aspirated configuration. The trace related to motored test at the same value of engine speed is also shown.

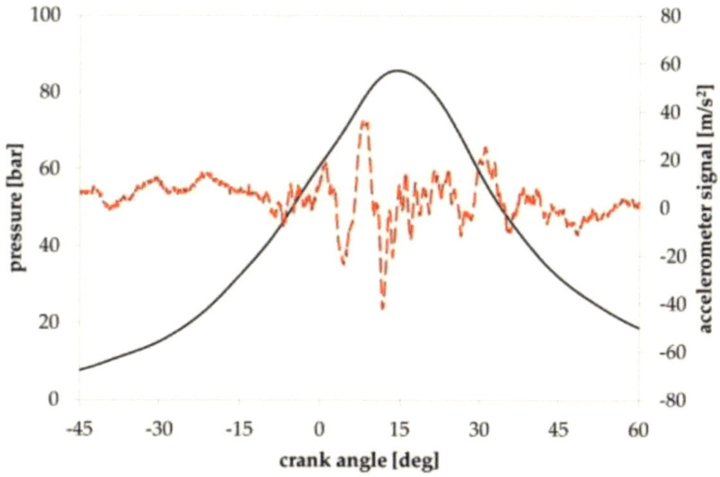

Figure 11. In-cylinder pressure and accelerometer signals at 4000 rpm, 100% load (turbocharged).

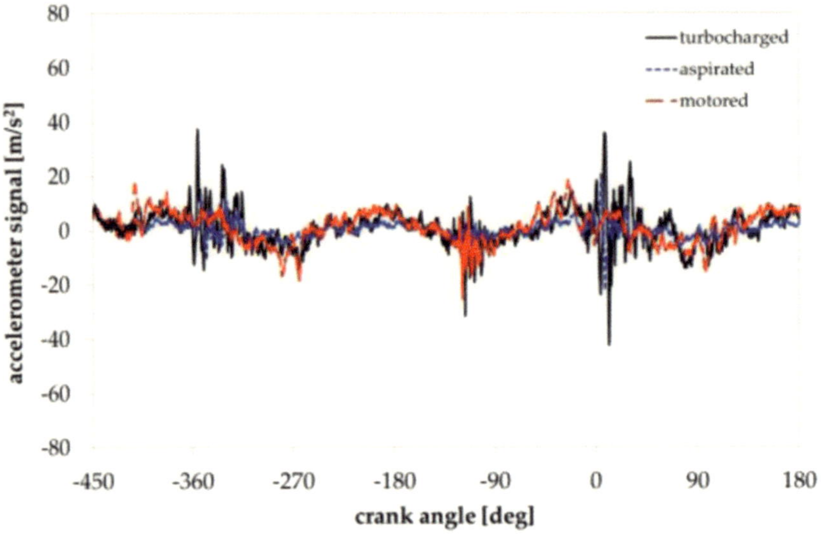

Figure 12. Accelerometer signals at 4000 rpm.

An analysis in the frequency domain of the data acquired with turbocharged configuration was performed and the frequency band in which in-cylinder pressure and accelerometer traces are highly correlated was evaluated. Starting from in-cylinder pressure data, ROHR, and CHR were computed (they are shown in **Figures 13** and **14**, respectively). In the plots, the data are superimposed on the vibration signal that was band-pass filtered according to the results of coherence function analysis. Circles are used to highlight in the accelerometer trace the crank angle values corresponding to the SOC and MFB50.

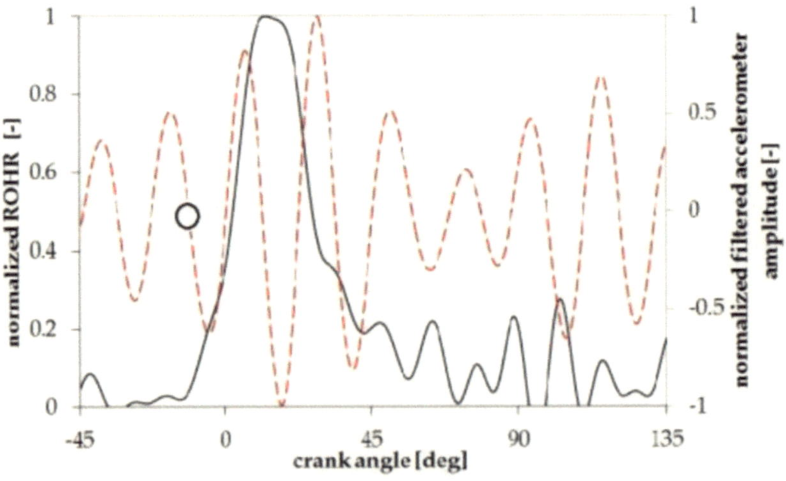

Figure 13. Normalized rate of heat release and filtered accelerometer trends at 4000 rpm, 100% load (turbocharged).

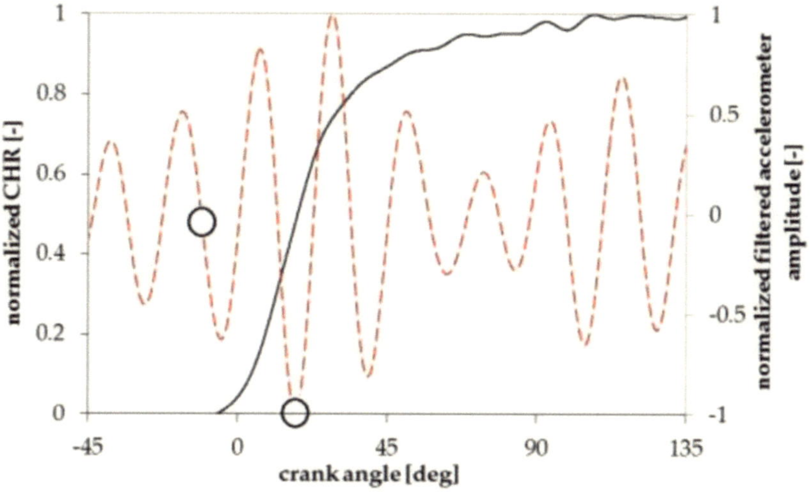

Figure 14. Normalized cumulated sum of rate of heat release and filtered accelerometer trends at 4000 rpm, 100% load (turbocharged).

Figures 15 and **16** show comprehensive plots of results obtained for naturally aspirated and turbocharged engine configuration, respectively.

In each figure, SOC and MFB50 are reported for 3600, 4000, and 4400 rpm, 60, 75, and 100 % load. Data on the *x*-axis show the crank angle value computed via CHR. Crank angle values in the *y*-axis were computed via filtered accelerometer trace.

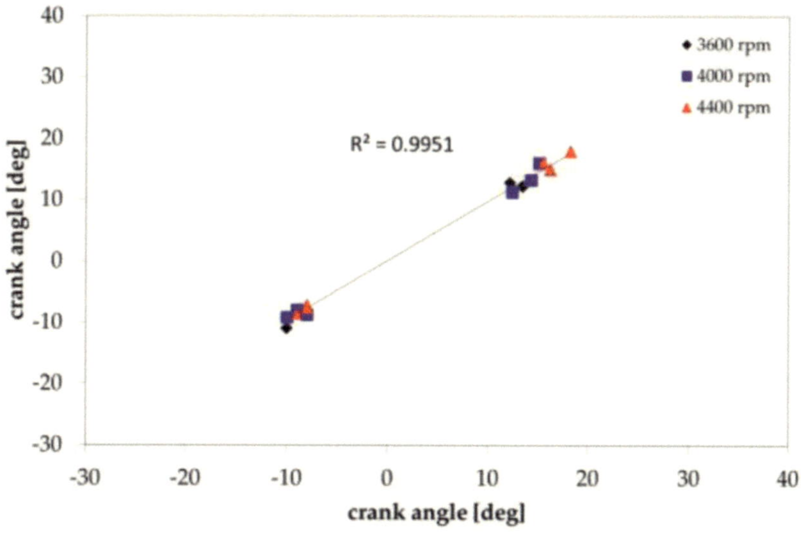

Figure 15. SOC and MFB50 for naturally aspirated engine configuration.

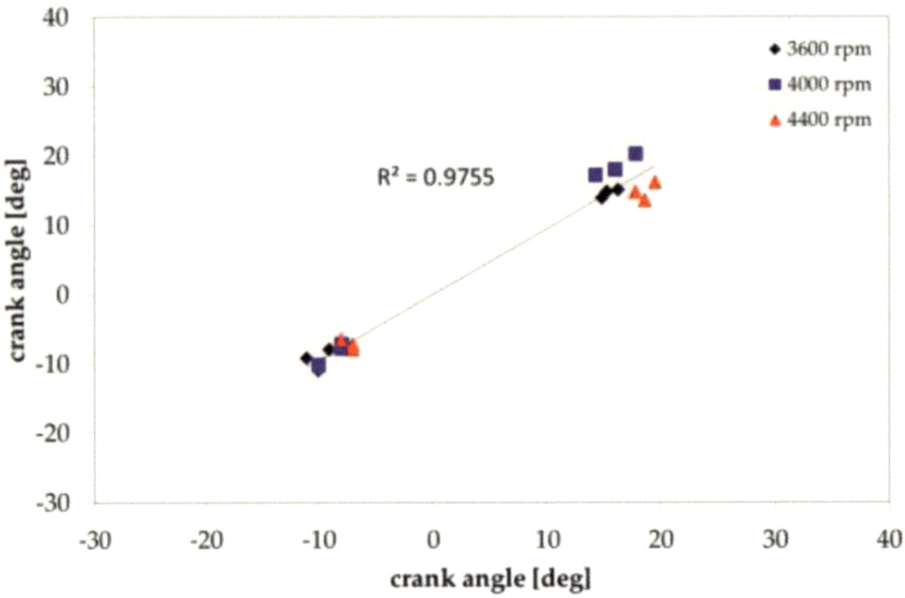

Figure 16. SOC and MFB50 for turbocharged engine.

In both plots, the interpolation lines and the corresponding R-squared values are shown (they are the square of the correlation coefficients).The obtained R values are in all cases very close to the unity, giving a measure of the very high reliability of the relationship between the combustion indicators estimated via accelerometer transducer and computed by direct in-cylinder pressure measurements.

4. Conclusion

A methodology was developed and validated, aimed at extracting from the signal of an accelerometer placed on a selected location of the engine block information about the combustion process in a diesel engine.

Experimentation was performed on a two-cylinder common rail diesel engine; two configurations were tested: naturally aspirated and turbocharged.

The analysis of the signals acquired in the engine complete operative field highlighted that it is always possible to select a frequency band in which in-cylinder pressure and engine block vibration signals are highly correlated. Such a band has demonstrated to be reliant on the engine speed value, whereas load condition has a weak effect on the frequency band, in agreement with results obtained during previous investigations.

The accelerometer signals were filtered in order to remove all the vibration components due to sources other than the combustion. The obtained combustion-related vibration contributions were used to evaluate indicators able to characterize the combustion development. The angular position of SOC and MFB50 was thus computed via processed accelerometer traces and compared to the same indicators evaluated via the heat release curve. The obtained data highlighted the high reliability of the methodology and indicated its prospective applicability in the real-time control of the engine management, in which the control algorithm manages the injection control unit based only on nonintrusive measurement. The comparison between combustion indicators evaluated only by means of the block vibration trend is compared to the optimal values stored in maps previously filled with data for each engine running conditions. The results of such a comparison are used as feedback signal to correct the injection settings.

Nomenclature

BTDC	Before top dead center
CDM	Code division multiplexing
CHR	Cumulative heat release
deg	Degree
main	Main injection
MFB50	Angular position at which half of the injected fuel is burnt
pil	Pilot injection
pre	Pre-injection
Q	Injected fuel
R	Correlation coefficient
ROHR	Rate of heat release
SOI	Angular position at which injection starts
SOC	Angular position at which combustion starts
TTL	Transistor-transistor logic

Author details

Ornella Chiavola*, Erasmo Recco and Giancarlo Chiatti

*Address all correspondence to: ornella.chiavola@uniroma3.it

Engineering Department, Roma Tre University, Rome, Italy

References

[1] Yang Z, Stobart R, Winward E. Online adjustment of start of injection and fuel rail pressure based on combustion process parameters of diesel engine. SAE Technical Paper 2013-01-0315. 2013. DOI: 10.4271/2013-01-0315

[2] Lujàn JM, Bermudez V, Guardiola C, Abbad A. A methodology for combustion detection in diesel engines through in-cylinder pressure derivate signal. Mechanical Systems and Signal Processing. 2010;**24**(7):2261-2275. DOI: 10.1016/j.ymssp.2009.12.012

[3] Lee S, Lee J, Lee S, Kim D, Lee Y, Yu S, Choi H. Study on reduction of diesel engine out emission through closed loop control based on the in-cylinder pressure with EGR model. SAE Technical Paper 2013-01-0322. 2013. DOI: 10.4271/2013-01-0322

[4] Ponti F, Ravaglioli V, Serra G, Stola F. Instantaneous engine speed measurement and processing for MFB50 evaluation. SAE International Journal of Engines. 2010;**2**(2):235-244. DOI: 10.4271/2009-01-2747

[5] Moro D, Cavina N, Ponti F. In-cylinder pressure reconstruction based on instantaneous engine speed signal. Journal of Engineering for Gas Turbines and Power. 2002;**124**:220-225. DOI: 10.1115/1.1391430

[6] Taglialatela F, Lavorgna M, Mancaruso E, Vaglieco BM. Determination of combustion parameters using engine crankshaft speed. Mechanical Systems and Signal Processing. 2013;**38**(2):628-633. DOI: 10.1016/j.ymssp.2012.12.009 DOI:10.1016/j.ymssp.2012.12.009 #doilink

[7] Desbazeille M, Randall RB, Guillet F, El Badaoui M, Hoisnard C. Model-based diagnosis of large diesel engines based on angular speed variations of the crankshaft. Mechanical Systems and Signal Processing. 2010;**24**:1529-1541. DOI: 10.1016/j.ymssp.2009.12.004; DOI: 10.1016/j.ymssp.2009.12.004#doilink

[8] Jiang J, Gu F, Gennish R, Moore DJ, Harris G, Ball AD. Monitoring of diesel engine combustions based on the acoustic source characterization of the exhaust system. Mechanical Systems and Signal Processing. 2008;**22**:1465-1480. DOI: 10.1016/j.ymssp.2007.12.003; DOI:10.1016/j.ymssp.2007.12.003#doilink

[9] Chiatti G, Chiavola O, Palmieri F, Piolo A. Diagnostic methodology for internal combustion diesel engines via noise radiation. Energy Conversion and Management. 2015;**89**:34-42. DOI: 10.1016/j.enconman.2014.09.055; DOI: 10.1016/j.enconman.2014.09.055#doilink

[10] Gu F, Li W, Ball AD, Leung AYT. The condition monitoring of diesel engines using acoustic measurements part 1: Acoustic characteristics of the engine and representation of the acoustic signals. SAE Technical Paper 2000-01-0730. 2000. DOI: 10.4271/2000-01-0730

[11] Ball AD, Gu F, Li W. The condition monitoring of diesel engines using acoustic measurements part 2: Fault detection and diagnosis. SAE Technical Paper 2000-01-0368. 2000. DOI: 10.4271/2000-01-0368

[12] Torii K. Method using multiple regression analysis to separate engine radiation noise into the contributions of combustion noise and mechanical noise in the time domain. SAE International Journal of Engines. 2014;**7**(3):1502-1513. DOI: 10.4271/2014-01-1678

[13] Kaul B, Lawler B, Zahdeh A. Engine diagnostics using acoustic emissions sensors. SAE International Journal of Engines. 2016;**9**(2):684-692. DOI: 10.4271/2016-01-0639

[14] Polonowski C, Mathur V, Naber J, Blough J. Accelerometer based sensing of combustion in a high speed HPCR diesel engine. SAE Technical Paper. 2007-01-0972. 2007. DOI: 10.4271/2007-01-0972

[15] Lee S, Lee Y, Lee S, Song H. Study on the correlation between the heat release rate and vibrations from a diesel engine block. SAE Technical Paper 2015-01-1673. 2015. DOI: 10.4271/2015-01-1673

[16] Jia L, Naber JD, Blough JR. Frequency response function adaptation for reconstruction of combustion signature in a 9-L diesel engine. Proceedings of the Institution of Mechanical Engineers, Part C: Journal of Mechanical Engineering Science. 2015;**229**(17):3071-3083. DOI: 10.1177/0954406215569256

[17] Jung I, Jin J, Won K, Yang S, Kyoungdoug M, Hoimyung C. Closed-loop control for diesel combustion noise using engine vibration signals. SAE Technical Paper 2015-01-2297. 2015. DOI:10.4271/2015-01-2297

[18] Arnone L, Boni M, Manelli S, Chiavola O, Conforto S, Recco E. Diesel engine combustion monitoring through block vibration signal analysis. SAE Technical Paper 2009-01-0765. 2009. DOI: 10.4271/2009-01-0765

[19] Chiatti G, Chiavola O, Recco E. Accelerometer measurements to optimize the injection strategy. SAE Technical Paper 2012-01-1341. 2012. DOI: 10.4271/2012-01-1341 234

[20] Chiatti G, Chiavola O, Recco E. Accelerometer signal to characterize the combustion development in multiple injection diesel engine. In: Proceedings of the ASME 2012 Internal Combustion Engine Division Spring Technical Conference; May 6-9, 2012; Torino, Italy. ICES2012-81235; pp. 659-666. DOI: 10.1115/ICES2012-81235

[21] Chiatti G, Chiavola O, Recco E. Combustion diagnosis via block vibration signal in common rail diesel engine. International Journal of Engine Research. 2014;**15**:654-663. DOI: 10.1177/1468087413512311

Chapter 6

Measurement of Exhaust Emissions under Actual Operating Conditions with the Use of PEMS: Review of Selected Vehicles

Jerzy Merkisz, Piotr Lijewski, Paweł Fuć,

Łukasz Rymaniak and Andrzej Ziółkowski

Additional information is available at the end of the chapter

http://dx.doi.org/10.5772/intechopen.70442

Abstract

This paper is a synthetic approach to real driving (RDE) from selected vehicles: light-duty vehicle (LDV), heavy-duty vehicle (HDV). The tests were performed with the portable emission measurement system (PEMS) equipment under actual traffic conditions. The paper discusses problems of measurement methodology and emission of CO, HC, NO_x, and PM. The performed investigation confirms that the main problem is the emission of NO_x and PM, which usually is higher than the emission level. The obtained results show that the RDE method is very complex, but is the only way to provide invaluable information on the actual on-road exhaust emissions, not obtainable under laboratory conditions. In recent years, methods of exhaust emission testing under actual operating conditions have been developing rapidly. New technologies for extra low engine emissions pose a new question: does emission testing in a standard laboratory reflects real life emissions of a vehicle in use? In order to answer this question, it is necessary to measure the vehicle in-use emissions. Today, we know that the engine operating conditions (engine load and speed) in laboratory tests are not compliant with the conditions of actual operation. That is why the results of such tests are so desirable.

Keywords: exhaust emission, HDV, PC, PEMS, RDE

1. Introduction

Exhaust emissions have long been the leading factor determining the improvement of powertrains and combustion engines. The technological advancement causes an increased emission of the greenhouse gases, CO_2 in particular, while one of the most important sources of its emission is the combustion of fuel in engines. Another aspect tightly related to the operation of combustion engines is engine exhaust emissions. Today, we know that the exhaust components generated by engines such as CO, HC, NO_x, and PM are hazardous to humans. The report published in 2012, International Agency for Research of Cancer (IARC), one of the branches of World Health Organization (WHO), informed that diesel exhaust gas causes cancer (IARC) [1]. Earlier, this exhaust gas was classified in a group containing factors referred to as probably carcinogenic. Upon analysis of the results of the most recent environmental research, the WHO scientists unanimously concluded that diesel exhaust gas causes cancer [2–4]. In the report of the US Environmental Protection Agency (EPA), following decades of research and laboratory tests on animals, it was confirmed that the particles are carcinogenic, and significantly contribute to the development of cancer, lung cancer in particular [5]. Therefore, carmakers treat the problem of exhaust emissions with priority.

Toxic exhaust emissions studies conducted in real operating conditions clearly show that the level of actual emissions from vehicles is greater than the test limit values [6]. Real driving emission (RDE) tests are very often used to optimize engine performance in terms of emissions and fuel economy. May et al. [7] point out that the emission of nitrogen oxides is greater in the RDE studies than that obtained in laboratory tests. Consequently, they recommend using the RDE test results for the optimization of engine control systems. Similar conclusions were reached in [8]. In this paper, the authors compared RDE emissions to simulation results using the COPERT simulating tool. Nitrogen oxides emissions in simulation tests were about 30% of the values obtained in the RDE tests. The authors of papers [9–11] point out that the important factors influencing the results of RDE tests are the conditions in which the research is conducted: traffic intensity, driver predisposition, or weather. Hence, the need to regulate the RDE testing procedures. It can therefore be concluded that due to the serious risk posed to human health and the need to develop techniques and methods for measuring toxic emissions, the RDE emissions tests are highly desirable.

The issue tightly related to the problem of exhaust emissions is the legislation controlling the exhaust emissions from engines and vehicles. Throughout the years, this legislation has evolved mainly toward reduction of the admissible levels of individual exhaust components and advancement of research methodology. Today, the procedures of exhaust emissions measurement include driving under actual operating conditions (RDE) using portable emission measurement system (PEMS). This type of research is becoming commonplace for all vehicle categories.

2. PEMS equipment

Currently applicable homologation legislation for heavy-duty vehicles (Euro VI) and the heralded proposal of future test procedures for this as well as other vehicle groups included the application of tests performed under actual driving conditions. This type of research, however, requires technologically advanced equipment (PEMS) that is increasingly often proposed by automotive measurement equipment manufacturers in their portfolio (AVL List GmbH, Horiba Ltd. and Sensors Inc.). This type of equipment can be used for testing machines and vehicles fueled with different fuels such as gasoline, diesel fuel, CNG, LPG, or oxygenated fuels. This also requires the application of special filters or exhaust gas diluters. Besides, the discussed equipment is characterized by high sampling frequency—a minimum of 1 Hz and reaching 500 Hz [e.g., high speed exhaust flow meter (EFM-HS)].

Due to the fast varying parameters of engines under actual conditions of operation and the advancement of aftertreatment systems, the equipment must be characterized by high measurement accuracy. The recording of ambient conditions is also necessary (pressure, temperature, humidity) as they have great impact on the measured values, hence additional corrective calculations are necessary. Therefore, the said measurement equipment often includes solutions (sensors, algorithms) that perform such procedures. What is more, the equipment must be characterized by low energy consumption, low weight and size, let alone high reliability. In order to perform a full analysis of the impact of a given type of powertrains and motion parameters of city buses on the environmental indexes, the following were utilized: SEMTECH DS (exhaust component concentration, oxygen content exhaust gas mass flow) as well as AVL MSS (used to determine the concentration of PM).

The presented measurement equipment is a unique set of analyzers allowing the determination of energy consumption and environmental performance of vehicles under actual conditions of operation:

- SEMTECH DS (**Figure 1**)—designed for the testing of gaseous exhaust components: NDIR: CO [%], CO_2 [%]; FID: THC [ppm]; NDUV: NO [ppm], NO_2 [ppm], and electrochemical O_2 [%];
- AVL MSS (Micro Soot Sensor—**Figure 2a**)—used to calculate the concentration of PM [mg/m^3] with the photo-acoustic method;
- TSI 3090 EEPS™ (Engine Exhaust Particle Sizer™ Spectrometer—**Figure 2b**)—allows determination of the size distribution of PM [nm];
- SEMTECH ECOSTAR (**Figure 3**)—allows measurement of both gaseous components and particulates (mass and number);
- AVL OTR (On The Road) OPACIMETER—used to test the exhaust gas opacity [%]; and
- SEMTECH LASAR—allows determination of the content of the exhaust gas including NH_3, N_2O, CH_4 [ppm].

The complementary analyzers are as follows: SEMTECH PPMD, SEMTECH LAM, AVL M.O.V.E., SEMTECH NMHC, AVL PARTICULATE COUNTER, TEXA NAVIGATOR TXT, and AVL INDIMICRO.

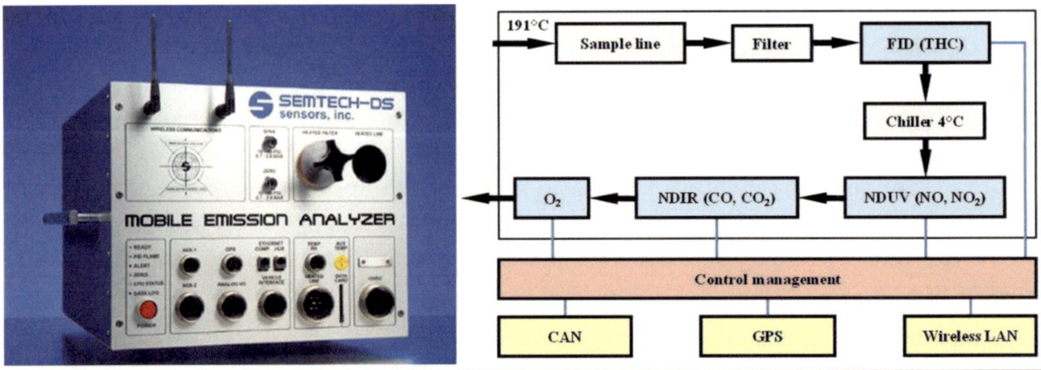

Parameter	Measurement method	Accuracy
Exhaust component concentration:		
CO	NDIR analyzer, measurement range 0–10%	±3%
HC	FID analyzer, measurement range 0–10 000 ppm	±2,5%
NO_x = (NO + NO_2)	NDUV analyzer, measurement range 0–3 000 ppm	±3%
CO_2	NDIR analyzer, measurement range 0–20%	±3%
O_2	Electrochemical analyzer, measurement range 0–21%	±1%
Exhaust gas flow	Mass flow rate	±2,5% range
	T_{max} to 700°C	±1% range

Figure 1. SEMTECH DS analyzer used for the measurement of exhaust gaseous components under actual conditions of operation [12].

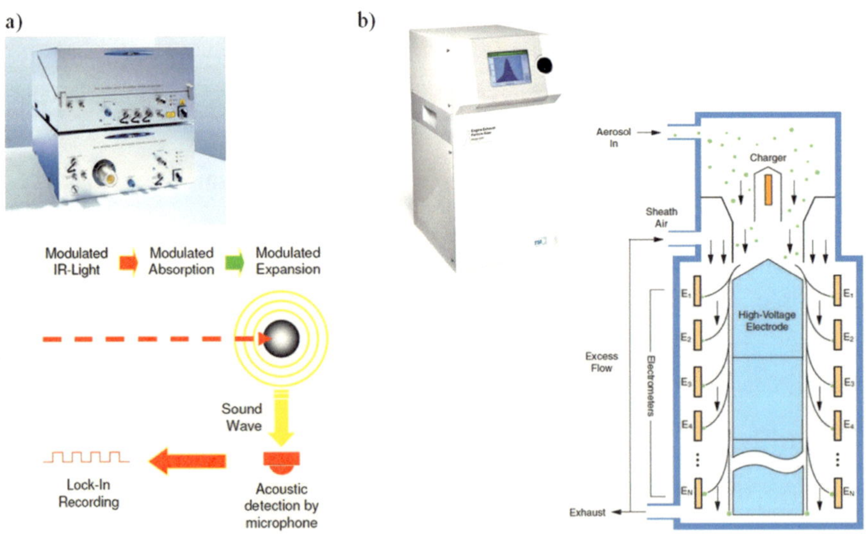

Figure 2. Equipment: (a) AVL MSS analyzers for the measurement of PM [13] and (b) TSI 3090 EEPS™ equipment for PM size distribution under actual conditions of operation [14].

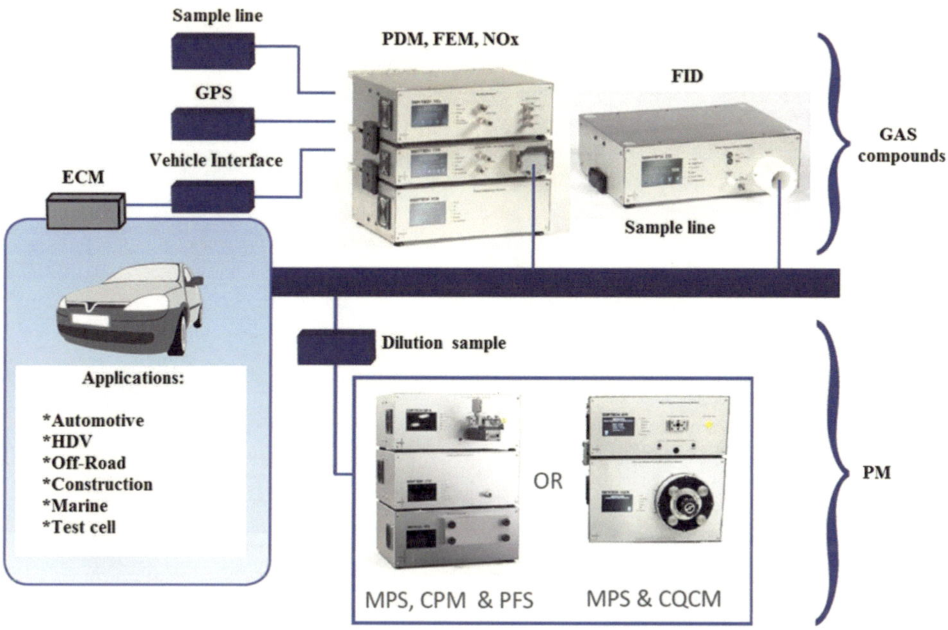

Figure 3. SEMTECH ECOSTAR analyzer for the measurement of gaseous exhaust components and particulate mass under actual conditions of operation [15].

3. Exhaust emissions legislation for light-duty vehicles

From the analysis of global exhaust emission standards for passenger cars (PC) and light-duty vehicles (LDV), we know that within the present decade, these standards will undergo modifications in all regions of the world (**Figure 4**). Due to the global reach of production and sales of motor vehicles by a variety of carmakers, a trend of unification of the test procedures is conspicuous. The exhaust emission limits are to remain varied and the introduction by a given legislator of additional tests will be allowed.

Currently homologation according to the ECE R83/06 directive describing the requirements for the obtainment of the Euro 5 and 6 standards. The details of the homologation legislation have been included in the EU (WE) 692/2008 regulation that is the amendment of regulation (WE) 715/2007. Item 1 article 3 of the (WE) 692/2008 regulation stipulates that in order to obtain a WE homologation in terms of exhaust emissions and the information related to vehicle repair and maintenance, the manufacturer confirms that the vehicles conform to the testing methods described in III–VIII, X–XII, XIV, and XVI of the said regulation. Aside from the said tests, the manufacturers are obliged to perform a procedure for conformity in operation, to check the operation of the on-board diagnostic system (OBD) and perform a measurement of the emission of CO_2 and fuel consumption. The regulation also includes the limits of individual exhaust components. For spark ignition engines of the Euro 6 standard, a limit of

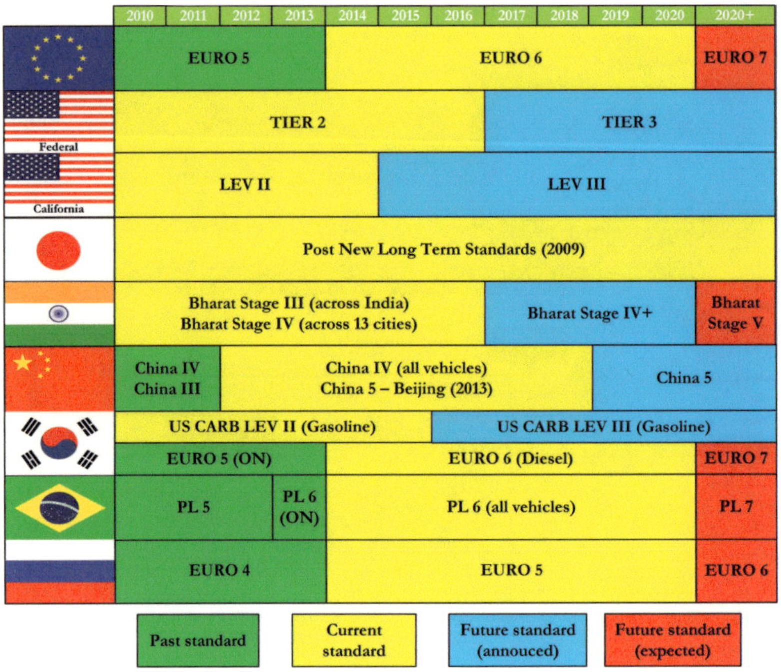

Figure 4. World emission standards 2010–2020 [16].

the particulate number (PN) was introduced for vehicles fitted with direct injected engines. Such regulations were not applicable in earlier standards.

Beside the exhaust emission limits, a very important aspect is the reduction of the emission of CO_2 from a fleet of vehicles. This emission is tantamount to fuel consumption. In 2007, European Commission (EC) proposed a 30% reduction of the emission of greenhouse gases in developed countries by 2020 and that the EU itself should undertake to reach at least a 20% reduction of these gases. Aside from the environmental aspects, the introduction of these regulations aims at accelerating and facilitating of the introduction of ultra low emission vehicles manufactured by the EU members. Regulation WE 443/2009 is applicable for the M1 vehicle category while WE 510/2011 is dedicated for the N1 category. Both legislations set the admissible limits of the road emission of CO_2, potential transition periods and deviations from these limits. The measurement of the CO_2 emission from both groups of vehicles is performed according to the methodology set forth in WE 715/2007 [17–19].

Prior to introducing the Euro 6 standard, an amendment of a series of regulations was planned. At the end of 2014, The European Commission passed to the European Parliament a motion (COM(2014) 28 final version) to amend the WE 715/2007 and WE 595/2009 regulation related to vehicle homologation procedures. The most important changes were as follows [16]:

1. increasing the gross vehicle weight from regulation WE 715/2007 (Euro 5 and 6) for the M_1, M_2, N_1, and N_2 vehicle categories from 2610 to 5000 kg;

2. changing the mass of CO_2 in the CoC mass of greenhouse gases as an equivalent of CO_2 and increasing or canceling the limit of the sum of hydrocarbons. Methane would then be construed as CO_2 equivalent;
3. introducing a limit of NO_2 emission into the supplement to the total NO_x limit. The maximum value of the emission must be determined based on the assessment of consequences;
4. supplementing of the tests (Type 6) with emission measurement of NO_x and NO_2, under low temperatures; and
5. authorizing the European Commission to update the limits of mass and number of particulates and the procedures of their measurement.

The most important amendment of the homologation procedures of PC and LDV vehicles is the replacement of the New European Driving Cycle (NEDC) with worldwide harmonized light vehicles test procedures (WLTP). They assume a global harmonization of driving cycles used for the testing of motor vehicles performed on chassis dynamometers. In ECE/TRANS/WP.29/2014/27 [20], a proposal has been presented for new test cycles that will be implemented in 2017. Three types of test cycles have been determined in this document, classified according to unit index of power (the ratio of effective power to the curb weight of the vehicle—PWr coefficient):

1. Class 1—low power output vehicles, PWr ≤ 22;
2. Class 2—vehicles in the range 22 < PWr ≤ 34; and
3. Class 3—high power output vehicles, PWr ≥ 34.

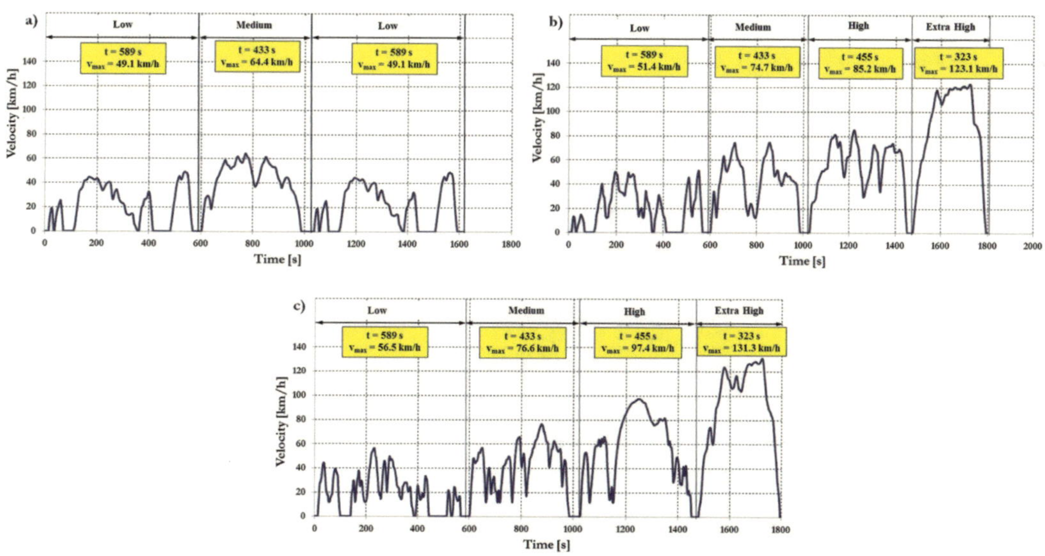

Figure 5. Curves of the WLTC driving cycle for vehicles: (a) Class 1, (b) Class 2, and (c) Class 3 [21].

For the Class 1 vehicles, the worldwide harmonized light-duty test cycle (WLTC) test is composed of three parts representing the driving conditions with two low and medium speed (**Figure 5a**). Its maximum value is 49.1 and 64.4 km/h, respectively. The average speed in the entire cycle is 33.3 km/h (counted without vehicle stationary) or 26.8 km/h including the vehicle stops that total 21.1% of the entire test duration. The test designed for Class 2 vehicles has an additional phase representing high speeds (**Figure 5b**). Its total time amounts to 1800 s and the vehicle covers a distance of 22.649 km. The values of the maximum and average speeds are different than those of the Class 1 vehicles. The most complex is the WLTC test for Class 3 vehicles (**Figure 5c**). It is composed of four phases. In the final part of the test, the vehicle develops a speed of 131.3 km/h [22]. In relation to the NEDC test, it is a 10% increase. The total WLTC test time for this class of vehicles is 1800 s. Comparing the WLTC test with the NEDC one, we can see a fundamental difference related to the velocity curve—the NEDC test is composed of repeated segments, while the WLTC tests have different velocity profiles that are a representation of the actual driving cycle. High variability of acceleration is a characteristic of these tests compared to the NEDC test of constant accelerations (**Table 1**).

	NEDC	WLTC Class 1	WLTC Class 2	WLTC Class 3
Time [s]	1180	1611	1800	1800
Distance [m]	11,023	11,428	22,649	23,262
Share of vehicle stationary [%]	33	21.1	15.8	13.4
Maximum speed [km/h]	120	64.4	123.1	131.4
Average speed [km/h]	33.6	26.8	50.4	51.8
Maximum acceleration [m/s^2]	1	0.76	0.96	1.58

Table 1. Characteristics of the NEDC and WLTC emission tests [20, 21].

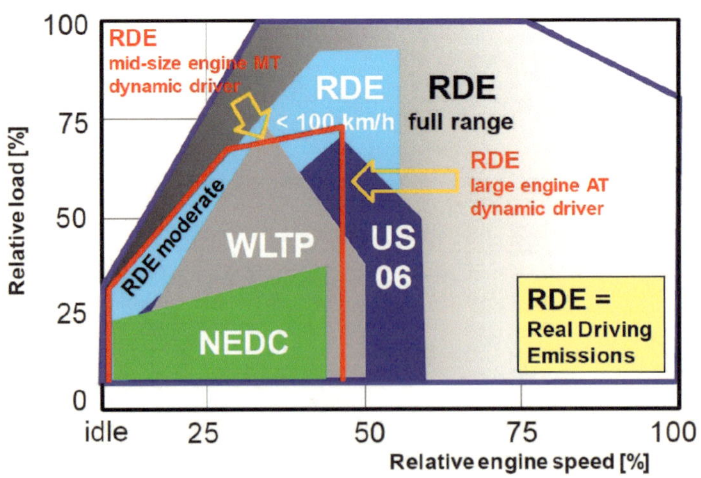

Figure 6. Engine parameters in different emission tests [22].

The work area of a combustion engine in the emission homologation tests is much smaller than the area occurring under actual operation (**Figure 6**). For this reason, an introduction of new methods of exhaust emissions measurements to the existing test procedures was proposed (real driving emissions). Portable emission measurement equipment will be used for this purpose (PEMS). On 10 March 2016, an EU Regulation—(UE) 2016/427 [23] was published containing a detailed description of the performance of these measurements for PC and LDV vehicles.

4. Tests of PC and LDV vehicles

4.1. Characteristics of the research objects and testing methods

Two vehicles were selected for the evaluation of the start-stop system. Vehicle A was fitted with a turbocharged gasoline engine of the displacement of 0.9 dm^3, fitted with a three-way catalytic converter (**Table 2** and **Figure 7**). The engine was characterized with a volumetric power output index of 70.8 kW/dm^3. Vehicle B was fitted with a diesel engine of the displacement of 3.0 dm^3. In this case, the volumetric power output index amounted to 58.7 kW/dm^3 and was lower by 17% than the index of vehicle A. The engine of this vehicle was fitted with a diesel oxidation catalyst (DOC) and a diesel particulate filter.

4.2. Testing methods

Exhaust emission tests (CO_2, NO_x, CO, and THC) were performed under real operating conditions of the vehicle in traffic in the Poznań city. The vehicle route during the tests has been shown in **Figure 8**.

The length of the route was 12.71 km. It was diversified and included a typical urban portion and an extra-urban portion where it was possible to drive at highway speeds (with a maximum speed of 120 km/h). The extra-urban portion was 5.5 km long. As shown in **Figure 8**, the length of the vehicle route during the road test was similar to that of the NEDC test [3]. The driving time in the road tests of approximately 1200 s was similar to that of the NEDC test.

Parameter	Vehicle A	Vehicle B
Type of ignition	Spark ignition	Compress ignition
Engine displacement	0.9 dm^3	3.0 dm^3
Cylinder number and arrangement	Straight—2	V—6
Maximum torque	145 N m @ 1800 rpm	550 N m @ 2000–2250 rpm
Volumetric power output index	70.8 kW/dm^3	58.7 kW/dm^3
Injection system	MPI	Common rail
Aspiration	Turbocharger	Turbocharger
Aftertreatment system	TWC	EGR, DPF, DOC
Type of transmission	Automatic	Automatic

Table 2. Characteristics of the tested objects.

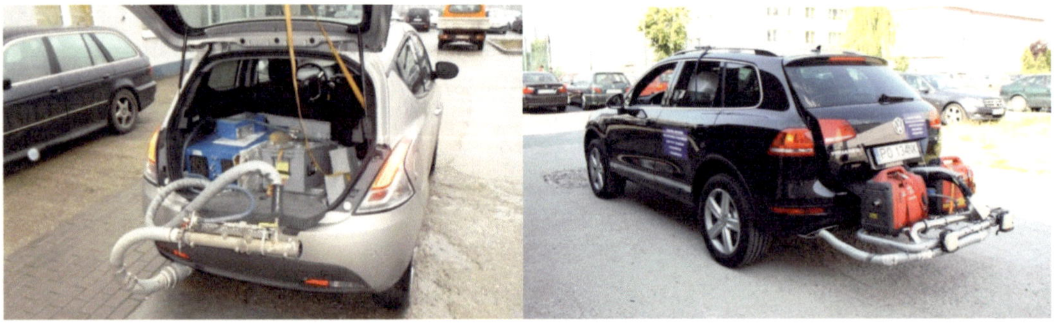

Figure 7. The tested objects ready for the on-road exhaust emissions tests.

4.3. Analysis of the exhaust emissions from LDV vehicles

In order to determine the efficiency of the start-stop system, the exhaust emission measurements were performed for the system in the enabled and disabled mode. Average speed was selected as a criterion decisive of the possibility of comparison of both vehicle drives. Its maximum relative difference was assumed on the level of 5%. For vehicle A, the relative speed difference was 3.5% and for vehicle B—5%.

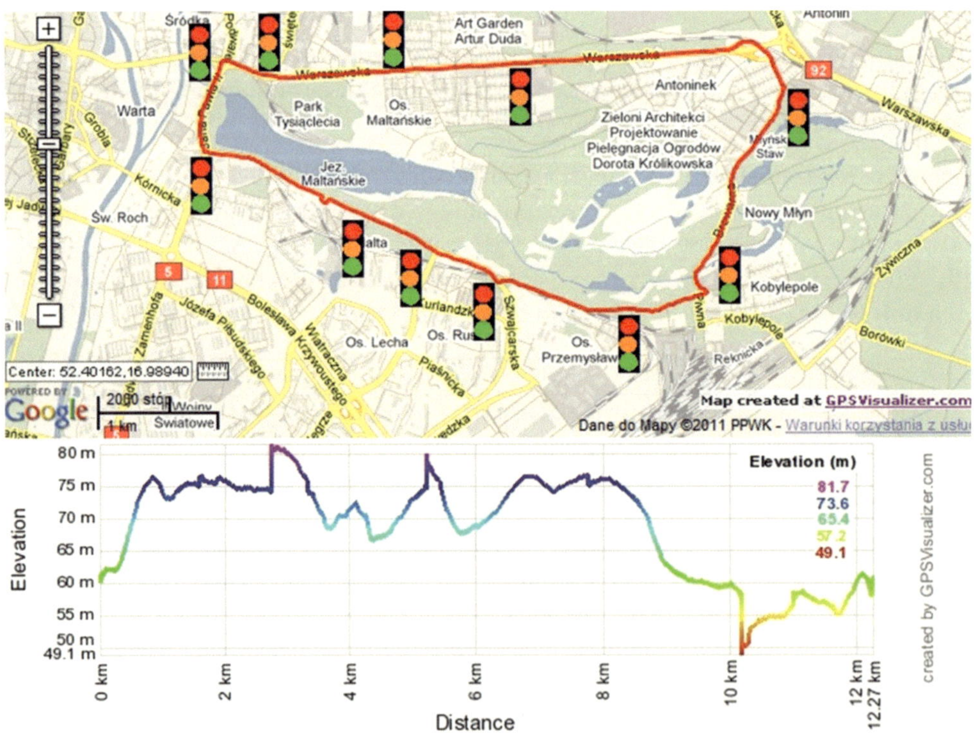

Figure 8. The road used for the exhaust emission testing (marked with red line) [created by GPSVisualizer.com].

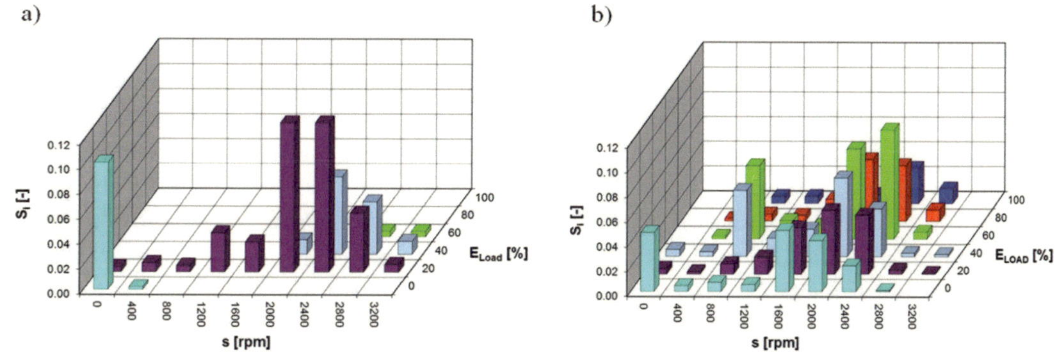

Figure 9. Characteristics of the operating time share referred to the engine (a) of vehicle A and (b) of vehicle B.

Based on the data recorded from the OBD system of the vehicles, the operating time share characteristics of the vehicle engines were made depending on the engine speed and torque (**Figure 9**). In the case of vehicle A, due to the start-stop system, as much as 11% of the driving time, the engine was off and for vehicle B it was 6%.

Figure 10 presents the changes in the exhaust emissions measured with the second-by-second resolution (CO_2, NO_x, CO, THC) and the engine speed on the example of vehicle B during a drive with the start-stop system enabled. Having analyzed the obtained courses, we have observed that the system switched off the engine seven times. The effect of this was obviously zero emission of CO_2 at that time. It has also been observed that in the first

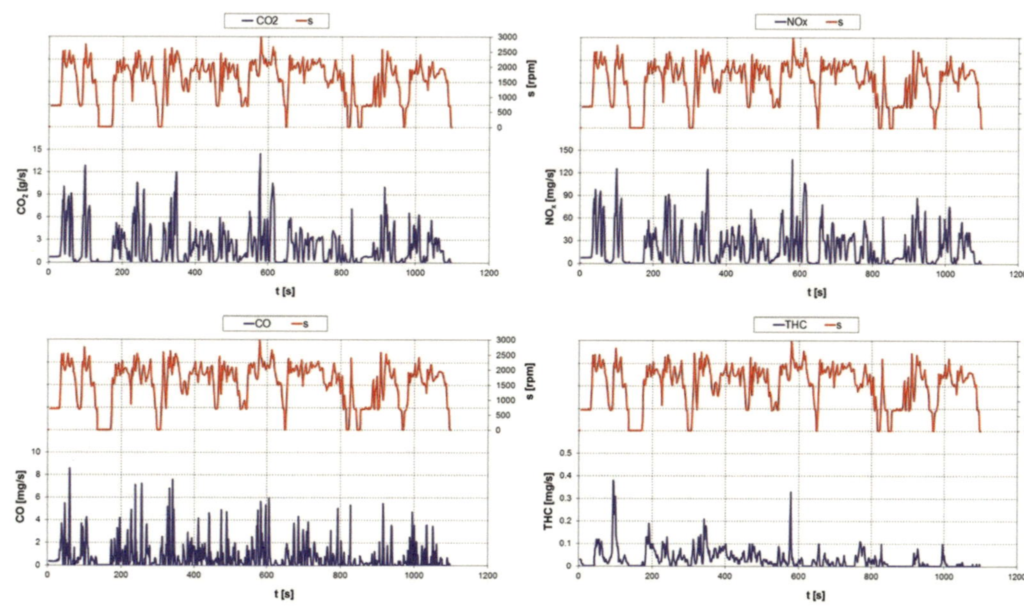

Figure 10. The courses of the emissions of CO_2, NO_x, CO, and THC measured with the second-by-second resolution and engine speed of vehicle B during a drive with the start-stop system enabled.

half of the test (0–600 s), the maximum values of the emission of CO_2 were lower than in the second part of the test. This depended on the characteristics of the test route—the second part of the test was a drive close to the extra-urban traffic conditions (a road portion of the entrance to the city). On this road portion, higher speeds were developed, which resulted in an increased energy demand of the engine, hence a growth in the emission of CO_2. In the case of the emission of NO_x and CO, a similar situation has been observed. The highest level of the emission of THC occurred in the first phase of the test. This most likely resulted from a cold engine start—lower temperature of the catalytic converter. We can deduct that the DOC catalyst in the beginning of the test did not reach the light-off temperature.

Based on the performed measurements, the on-road emissions of CO_2, NO_x, CO, THC, HC + NO_x as well as gas mileage were determined. The obtained values were compared to the limits set forth in the Euro 5 standard (**Figure 11**). The authors also determined the efficiency of the applied start-stop system. For vehicle A, the enabling of the system resulted in a reduction of:

- the emission of CO_2 by 7%;
- the emission of NO_x by 13%;
- the emission of CO by 12%;
- the emission of THC by 15%; and
- gas mileage by 9%.

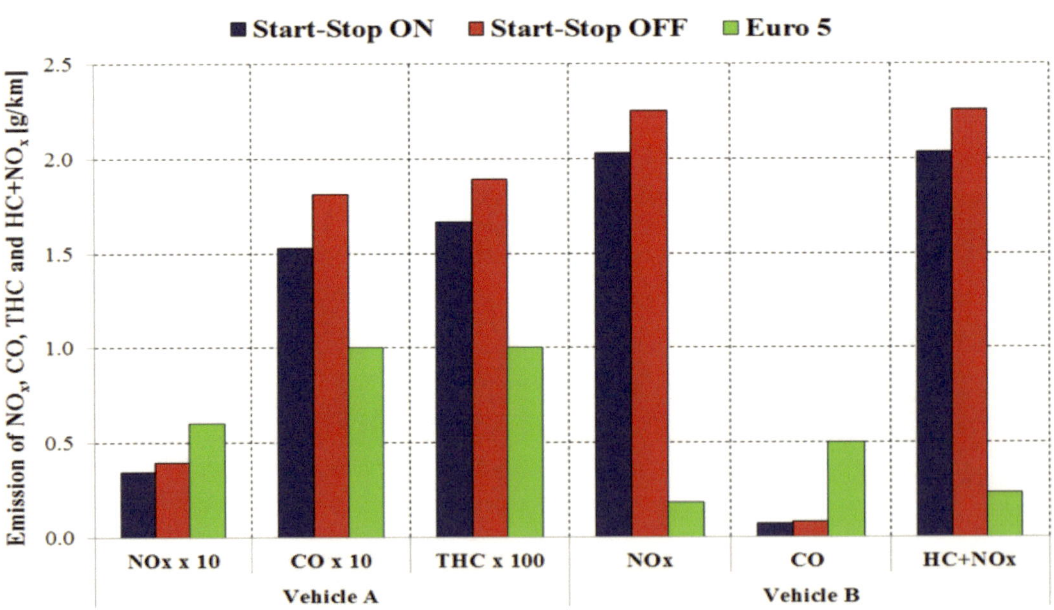

Figure 11. The obtained on-road emissions of NO_x, CO, THC, and HC + NO_x referred to the limits determined in the Euro 5 standard.

For vehicle B, a reduction has been recorded of:

- the emission of CO_2 by 11%;
- the emission of NO_x by 10%;
- the emission of CO by 13%;
- the emission of HC + NO_x by 10%; and
- gas mileage by 12%.

The obtained results confirm the efficiency of the application of the start-stop system. It is worth emphasizing that the presented results reflect the actual benefits from the application of this system as the tests were performed in actual traffic; hence, certain operating factors having impact on the exhaust emissions and fuel consumption were taken into account—factors that are not always considered during laboratory tests.

5. Exhaust emissions legislation for engines of HDV vehicles

For the HDV group of vehicles, due to their specific design characterized by significant size and power outputs, homologation in terms of unit emissions is performed for the engine alone on engine test beds. For all Euro standards, performing tests at preset points of work in stationary tests is necessary. Since 2000, the Euro III standard is applicable for the homologated vehicles. It is based on: European transient cycle (ETC), European stationary cycle (ESC), and European load response (ELR). The same procedures are applicable for the Euro IV, V, and EEV standards, but the exhaust opacity test for heavy-duty vehicles is performed in the ELR test. The boundary values of the exhaust emissions for diesel engines have been reduced with the introduction of further directives and regulations (**Figure 12**). A relative reduction of the unit emissions of individual exhaust components was: CO—67%, HC—88%, NO_x—95%, and particulate mass—98%, respectively. For exhaust opacity, limits were applicable throughout the Euro III–V standards. Following the technological advancement of combustion engines and aftertreatment systems, an introduction of boundary values [particle number (PN)] in the Euro VI became necessary. Along with the introduction of the Euro III standard, an European transient cycle (ETC) became applicable in which the combustion engine was tested for dynamic operating parameters (**Figure 13**).

In the Euro VI standard, in relation to the previous regulations, requirements related to the diesel engine have been more extensively defined. The latest regulations do not allow for the additional division into regular standards and EEV.

During the development of the transient *World harmonized transient cycle* (WHTC) and stationary *World harmonized stationary cycle* (WHSC), road test results performed in selected EU member states, Japan and USA were taken into account. This aimed at approximating the test bed measurement cycles to the actual conditions of operation worldwide. The

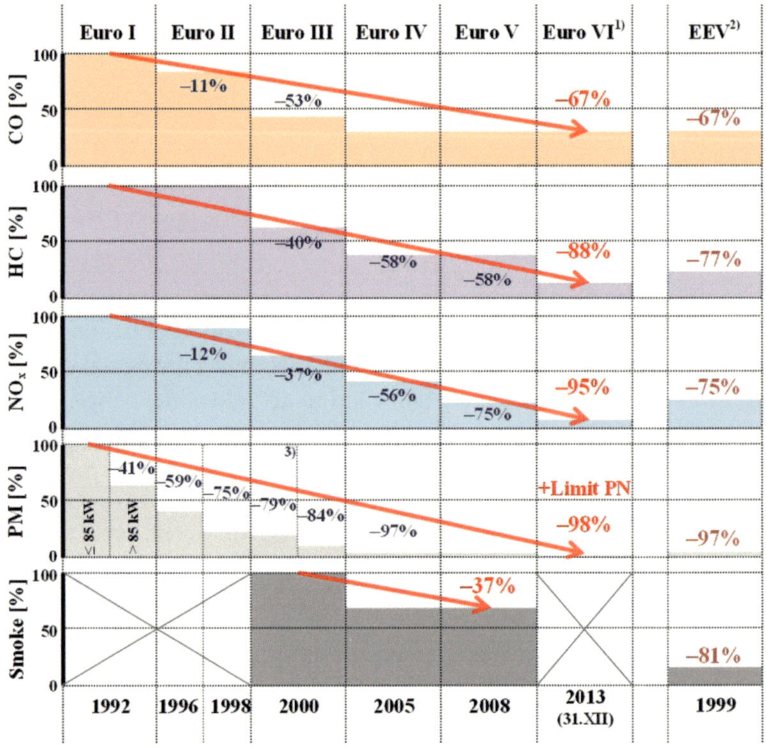

1) for the Euro VI standard NH_3 limits were also introduced

2) the standard was applicable for Euro III–Euro V in the ESC test

3) for engines of the displacement lower than 0.75 dm³/cyl. and engine speed greater than 3000 rpm

Figure 12. Relative reduction of the admissible unit emission values related to the subsequent Euro standards for stationary tests [24, 25].

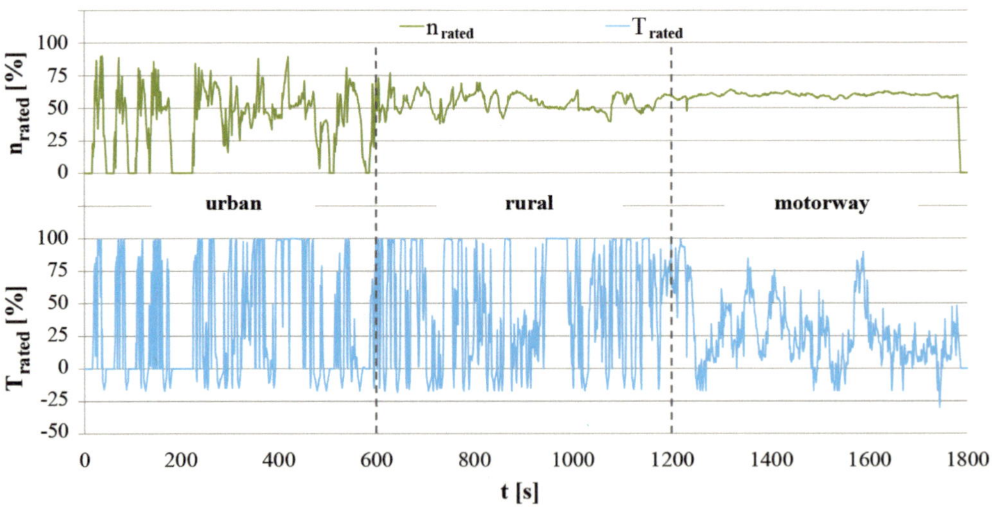

Figure 13. The engine torque and speed curves during the ETC test [26].

previously applied tests were developed based on the European conditions exclusively. Detailed data on the realization of the tests are included in the *Global technical regulations* (GTR) no. 4 developed by UNECE [27]. Similarly to the transient cycle applicable for the Euro V standard, the WHTC trials can be realized on both chassis dynamometers and engine dynamometers, yet for the homologation, the latter option applies (**Figure 14**). There are three parts to the tests—urban driving, extra-urban driving, and expressway driving. The length of the trial is 1800 s. The length of the subsequent phases of the tests is 900 s, 481 s, and 419 s, respectively.

The WHSC test was designed to measure the exhaust emissions under steady state conditions and is composed of 13 phases, similarly to the ESC test (**Figure 15**). The main difference between the said cycles is different engine speeds (the latter standard provides for six main engine speeds) and the application of different weight coefficients, resulting from the durations for each point of work. Besides, the measurement at idle occurs twice. It is noteworthy that the average values of n and T are lower than in the ESC test, which indicates that the engine is under smaller load.

In the Euro V and Euro VI standards, mileages and periods of operation are given corresponding to a regular life cycle of a heavy-duty vehicle. Minimum mileages are also provided for the durability trials depending on the category and gross vehicle weight of the tested object. The legislation is supplemented with data related to the determination of coefficients of deterioration (more precise and unambiguous definitions have been included in the Euro VI standard) [18]. Important changes in the regulations were introduced in terms of the scope of supervision of the conformity in operation—it is now necessary to perform exhaust emission measurements under actual traffic conditions using PEMS. Earlier, only engines (removed from the vehicle) were tested.

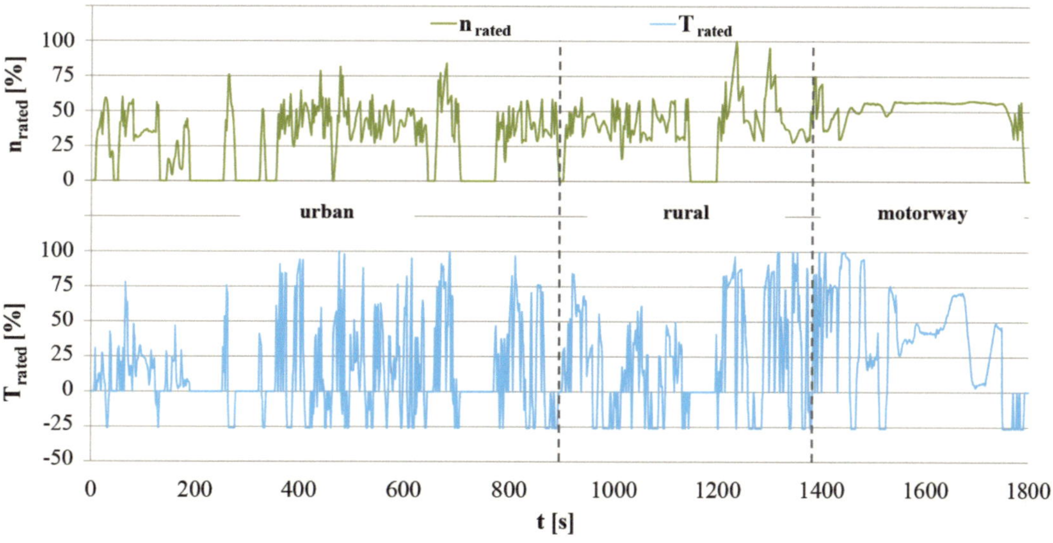

Figure 14. The engine torque and speed curves during the WHTC test [27].

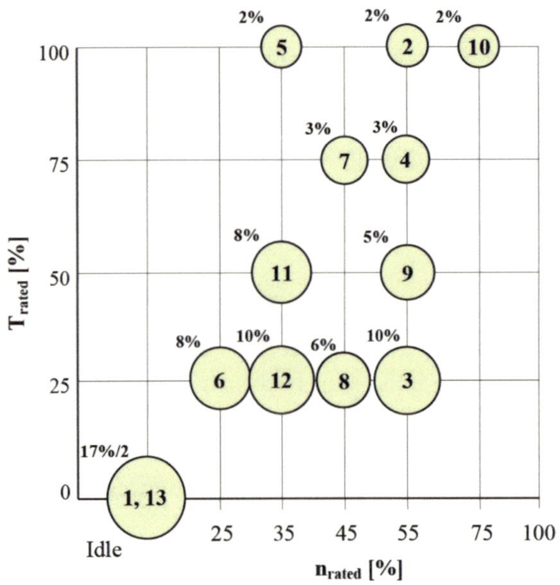

Phase	Time covering the interval [s]
0	–
1	210
2	50
3	250
4	75
5	50
6	200
7	75
8	150
9	125
10	50
11	200
12	250
13	210

Figure 15. The course of the harmonized WHSC test (numbers 1 through 13 represent the order of phase completion; percentage values at subsequent phases indicate their weights) [27].

Regulation UE 582/2011 (schedule II) includes detailed requirements related to the determination of conformity in operation of the engines or vehicles. Out of the most vital information, it should be mentioned that the measurements must be performed on public roads of the UE member states, using typical driving styles and loads. This means that the tests are performed for standard (most frequently occurring) conditions of operation. During the test procedure, it is also important that the driver has sufficient skills and is properly trained to use the vehicle, preferably a person who uses this particular vehicle on a daily basis. If the tests cannot be performed under standard operating conditions, it is possible to use alternative routes. In light of the shortage of information on representative vehicle load, a replacement load is applied ensuring 50–60% of the maximum load.

Upon first registration of a complete vehicle fitted with a combustion engine from the homologated group of engines, the manufacturer must perform tests in operation within 18 months reaching a mileage of at least 25,000 km [28]. According to [29], the measurements must be repeated periodically, at least every 24 months throughout typical vehicle life cycle. The test route must include urban roads (speed range: 0–13.89 m/s), extra-urban roads (13.89–20.83 m/s), and expressways (in excess of 20.83 m/s). In justified cases, the order of the test routes may be changed. The shares of the drives under individual conditions depend on the category of the tested vehicle (**Table 3**). They are determined with the accuracy of ±5%, due to the actual traffic conditions that are hard to predict. It is very important that in the realized test route, five times the work performed during the WHTC test is obtained or five times the reference mass of CO_2 from the same test is reached.

The coefficients of conformity are determined for measurement windows identified with two methods: based on the mass of CO_2 or total work performed by the engine. In the legislation

Category	Share of operating conditions [%]		
	Urban	Rural	Motorway
M_1, N_1	45	25	30
M_2, M_3 Class I, II & A	70	30	0
M_2, M_3 other	45	25	30
N_2	45	25	30
N_3	20	25	55

Table 3. Share of the operating conditions during the conformity in operation tests (HDV) [28].

process, in order to issue a decision, only the second variant should be performed. The tests include the unit emission of gaseous exhaust components: CO and THC (for diesel engines), NMHC and CH_4 (for spark ignition engines) as well as NO_x (for diesel and spark ignition engines). Currently, the mass and number of particulates is not taken into account. Besides, it is necessary to measure the exhaust gas mass flow, engine parameters, vehicle speed, ambient conditions, etc.

The assessment of the unit exhaust emissions is made using variable averaging windows (**Figure 16**). Their determination consists in obtaining the mass rate of the exhaust emissions for subsets of a complete data set, whose length is determined so that they correspond to the mass of CO_2 generated by the engine or work measured under transient conditions on a test bed (WHTC).

The condition for accepting the measurement window as valid is fulfilling the requirement of reaching an average power output exceeding 20% $N_{e\,max}$ in that window. In the entire test, the percentage of measurement windows must be 50% or more. If this is not obtained, the data

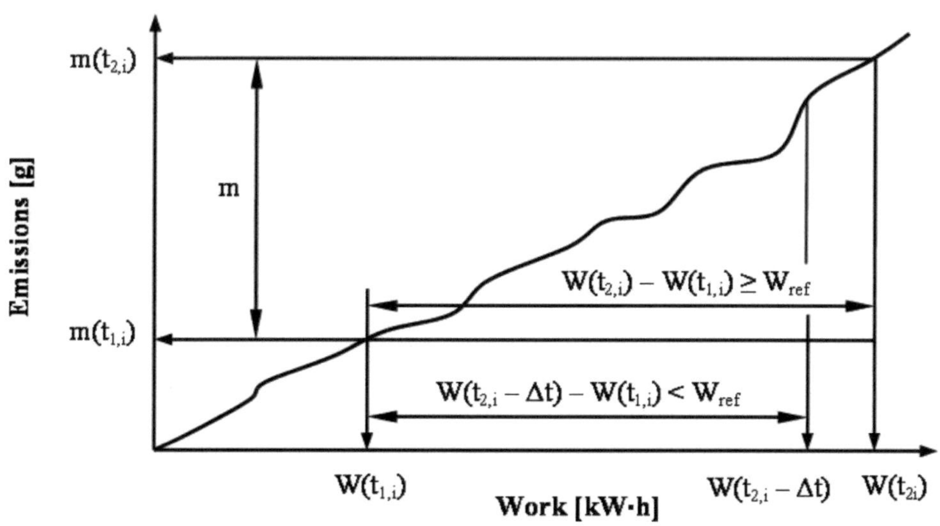

Figure 16. Determination of the measurement windows in the method based on reference work [28].

evaluation is repeated applying lower power output thresholds. The reduction is made with 1% resolution, maximum 15% $N_{e\,max}$. Lower value renders the results invalid.

The coefficient of conformity in operation in terms of exhaust emissions *conformity factor* (CF) is determined in all windows for each analyzed exhaust component as per Eq. (1). In order to render the evaluation in a given averaging interval positive, the determined coefficients cannot be greater than 1.5. The vehicle is considered compliant if 90% of the calculated CF values meet this criterion.

$$CF = \frac{e_j}{L_j} \quad (1)$$

where CF—coefficient of conformity in a given averaging window; L_j—admissible emission of a *j*th component in the WHTC test [mg/(kW h)].

The American *United States Environmental Protection Agency* (US EPA) has proposed a test serving the purpose of controlling the exhaust emissions from heavy-duty vehicles under non-test conditions *not-to-exceed* (NTE) that could be applied during the assessment of the actual environmental indexes. The NTE requirements were introduced in 1998 as an ordinance with the consent of the HDV engine manufacturers [25]. The test stringent requirements were gradually extended to other engine categories. As an assumption, the limits and procedures of the test performance were developed as an additional confirmation that exhaust emissions are in conformity with the legislation in the entire range of engine speeds and loads. The NTE test area of an example engine has been shown in **Figure 17**.

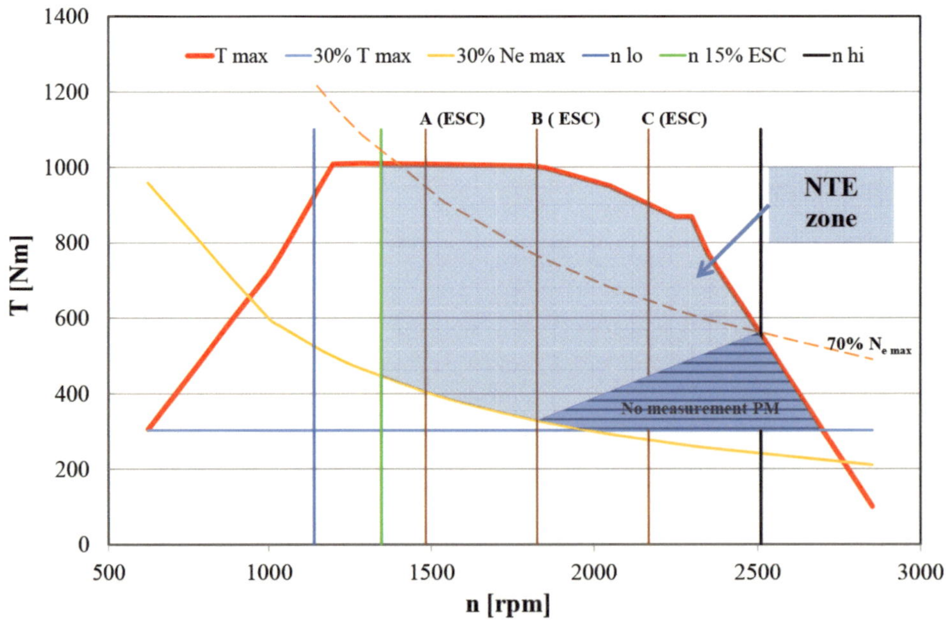

Figure 17. The NTE test area for the engine of a heavy-duty vehicle.

The test procedure does not determine any specific driving cycle, but it contains guidelines related to the determination of the area for which the emission is measured. The tests can be performed under any operating conditions. It is vital that in a certain range, the engine operates for at least 30 s in the control area. In this period, the values of engine speed and load may not exceed a predetermined area. The obtained values of exhaust emissions are averaged and in the final stage compared to the applicable NTE emission limits. In order to determine the control area, it is necessary to determine the basic engine characteristics and meet the following conditions [25]:

- the minimum engine speed (engine A) is analogical to the engine speeds of the European stationary cycle (ESC) test, just as is in the case of engine B;
- engine load is equal or greater than 30% of the maximum engine torque;
- all points of work in which the engine develops power lower than 30% of the maximum power are excluded from the NTE area; and
- the engine manufacturer may apply for an exclusion from the NTE area of engine speeds and loads for which the unit fuel consumption is greater than 5% of the minimum fuel consumption, if he expects that the engine will not work in these areas during regular operation. This does not apply to engines coupled with automatic transmissions of a specified number of speeds and vehicles with manual transmissions.

Due to the obtained maximum engine speeds (2400 rpm), there are small differences in defining the area excluded from the measurement of particulate matter.

6. Testing heavy-duty vehicles

6.1. The route

For the tests, the authors selected a road portion of the length of 27 km (**Figure 18**). The road portion well characterizes the operation of vehicles of the GVW exceeding 16,000 kg (long haulage) in the area where the measurements were carried out. The test route started and ended in the industrial zone (point A) where a production facility is located at which approximately 50 heavy-duty vehicles are handled daily. The test road portion can be divided into two parts: a drive on urban roads (portion A–B) and national and regional roads. The drive on national or regional roads depends on the driving direction from/to the entrance to the A2 expressway (Koło) (point D). In the case of driving to the "Koło" expressway entrance, the route went through points B–C and C–D. In the reverse situation, i.e., exiting the expressway and driving to the production facility via bypasses: points D–C and C–B (on the D–C road portion heavy-duty trucks of the GVW in excess of 7000 kg are not permitted.) The above route can be deemed representative of the national transport and logistic infrastructure—representing the road infrastructure and the distribution of production facilities in small and medium-sized towns.

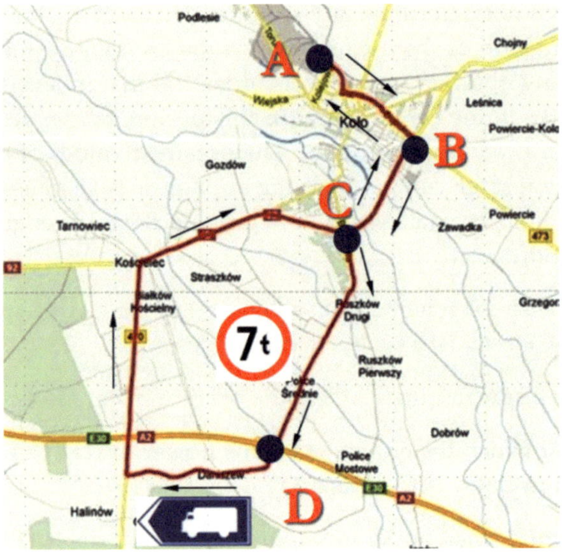

Figure 18. The measurement road portion used in the on-road emission tests [made based on GPSVisualiser.com].

6.2. Research objects

For the research, the authors used two heavy-duty trucks (road tractors with semi-trailers) loaded with a cargo of 20,000 and 24,800 kg (**Figure 19**). The first of the objects was fitted with a 309 kW (420 KM) Euro III engine. The other object had a V8 412 kW (560 KM) Euro V engine. Both vehicles were fitted with an automatic transmission (**Table 4**) of the 12+1 configuration. The second vehicle was also fitted with a driver monitoring system. By a continuous analysis of signals from a series of sensors, the system provides real time suggestions, and upon end of trip generates a report on the driving style. The suggestions and the evaluation are presented on a display and have four categories: driving uphill, predicting, braking, and gearshifts. The idea behind the system is to continuously improve the driving skills in terms of fuel consumption and proper use of modern solutions such as: automatic transmission, retarders, or *electronic braking system* (EBS).

a) b)

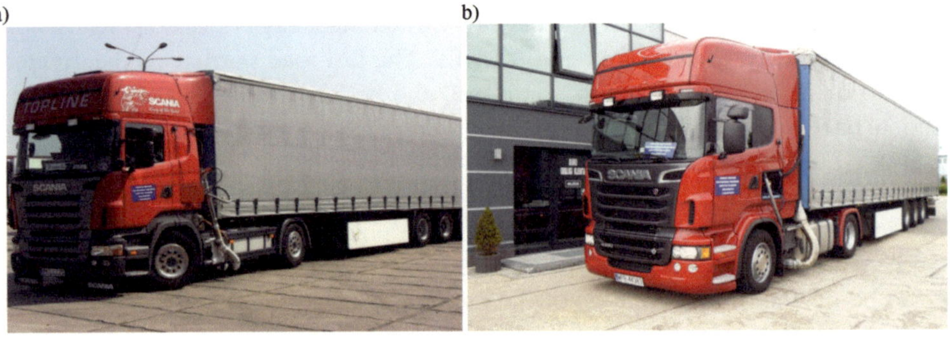

Figure 19. Research objects during the on-road emission tests: (a) vehicle A and (b) vehicle B.

Parameter	Vehicle A	Vehicle B
Displacement	11.7 dm^3	15.6 dm^3
Number of cylinders/arrangement	6/straight	8/V8
Maximum power output	309 kW @ 1900 rpm	412 kW @ 1900 rpm
Maximum torque	2100 N m @ 1000–1350 rpm	2700 N m @ 1000–1400 rpm
Unit power output index	8.3 kW/t	10.3 kW/t
Emission standard	Euro III	Euro V
Exhaust gas aftertreatment	N/A	SCR
Transmission	Automatic 12 + 1	Automatic 12 + 1
Driver support system	N/A	SDS
Tractor axle configuration	4 × 2	4 × 2
Curb weight including trailer	15,000 kg	15,200 kg
Cargo weight	20,000 kg	24,800 kg
Type of cargo	Big-Bag	Steel
Type of trailer	Canopy	Canopy

Table 4. Characteristics of vehicles used for the tests.

6.3. The exhaust emission correction coefficient

Because the authors could not perform the measurements on two heavy-duty vehicles of the same exhaust emissions standard, the tests were carried out for Euro III and Euro V compliant vehicles. In order to compare the obtained values of the CO emission, the authors decided to define a dimensionless emission correction coefficient C_i [25]:

$$C_i = \frac{e_{\text{EuroV}}}{e_{\text{EuroIII}}} \tag{2}$$

where C_i—correction coefficient of an *i*th component; e_i, Euro V—a limit of unit emission of a given component in the Euro V standard [g/(kWh)]; e_i, Euro V—a limit of unit emission of a given component in the Euro III standard [g/(kWh)]. Determined values of C_i coefficient are shown in **Table 5**.

	Euro III [g/(kWh)]	Euro V [g/(kWh)]	Coefficient C_i [–]
CO	2.10	1.50	0.72
NO_x	5.00	2.00	0.40
PM	0.1	0.02	0.20

Table 5. Values of the emissions limit and C_i coefficient.

6.4. Analysis of the vehicle driving profiles

In the first, urban part, significant differences in the driving profiles of both vehicles were recorded. Vehicle A had a higher speed than vehicle B (**Figure 20** and **Table 6**). This was caused by higher traffic congestion during the test run of vehicle B. In the rural part, both driving profiles were similar. Vehicle A, during the entire run had a lower average speed (by 5%) than vehicle B. From the analysis of the maximum and average acceleration in the acceleration phase, it results that vehicle B was more dynamic because in both cases its values were higher by 49 and 19%, respectively. The second-by-second emission of CO, NO_x, and PM of vehicle A was multiplied by index C_i and then compared with the course recorded for vehicle B.

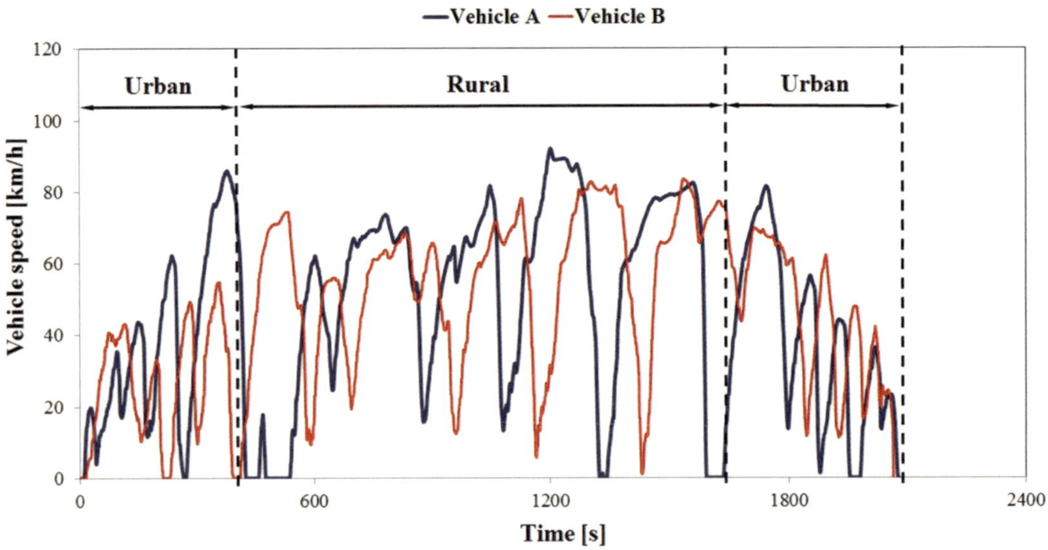

Figure 20. Speed profiles of the tested vehicles obtained during the on-road tests described with function $V = f(t)$.

Parameter	Unit	Vehicle A	Vehicle B	Percentage ratio vehicle A/ vehicle B [%]
Distance, s	km	26.57	26.88	98.84
Maximum speed, V_{max}	km/h	92.00	84.52	108.14
Average speed, V_{ave}	km/h	45.73	48.56	94.17
Minimum acceleration, a_{min}	m/s²	−2.93	−2.80	104.64
Maximum acceleration, a_{max}	m/s²	1.28	2.53	50.60
Average acceleration in phase of the ramp-up, $a_{śr}$	m/s²	0.21	0.26	80.77

Table 6. Parameters characterizing the test runs of both vehicles during the on-road tests.

Analyzing the second-by-second emission of NO_x, the authors observed that in the first, urban phase vehicle B had higher values of this emission than vehicle A (**Figure 21**). In the further part of the test, this trend changed and vehicle A had higher emissions. Such a situation was caused by the selective catalytic reduction (SCR) system responsible for the control of the NO_x emission, fitted in the exhaust of vehicle B. In the first phase of the test, the SCR system was most likely inhibited, as the tests for both vehicles were initiated from a cold start (a cold start is to be construed in this case as starting the engine at an ambient temperature of over 20°C) and under these conditions, the exhaust gas temperature is too low for the NO_x reduction to take place if a 32.5% water solution of urea is applied. Upon stabilization of the engine thermal state, a growth of the exhaust gas temperature takes place; thus, generating proper conditions for the NO_x reduction. The temperature of the exhaust gas also influences the conversion rate of the SCR catalytic converter where the said reactions take place. In standard SCR converters, the highest conversion rate occurs for 250–400°C. Under such conditions, the SCR control system initiates injection of a 32.5% solution of urea into the vehicle exhaust system, from which, following a series of reactions, ammonia is generated and used in the selective reduction of NO_x. From the recorded course of the second-by-second emission of NO_x, it results that the SCR system had obtained the highest conversion rate after 600 s of the test run, and that vehicle B obtained much lower values of this emission than in the initial phase of the test. In this part of the test, vehicle B also had lower emission of NO_x compared to vehicle A.

Next, based on the carbon balance method [30], the gas mileage for both vehicles was determined (in this method, the on-road emission of CO_2, CO, and HC is taken into account). The HC part has been omitted due to relatively low values of the on-road emission of this component by heavy-duty trucks remaining within the margin of measurement error. **Figure 22** presents the comparison of the on-road emissions of CO, NO_x, CO_2, PM, and the gas mileage. In all cases, vehicle B obtained lower values and a higher gas mileage. It is noteworthy that it had a significant increase

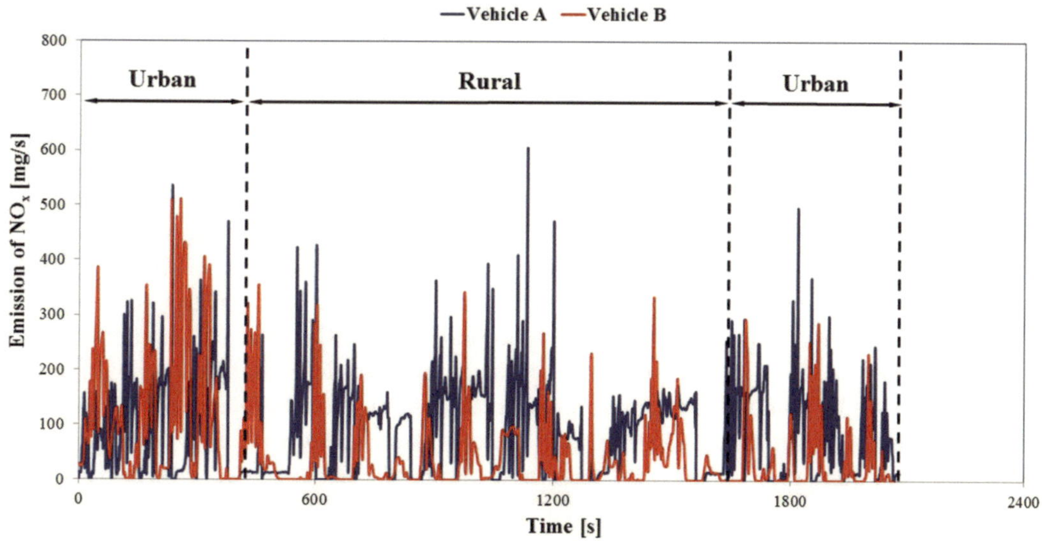

Figure 21. The tracing of the second-by-second emission of NO_x for both vehicles obtained during the on-road tests.

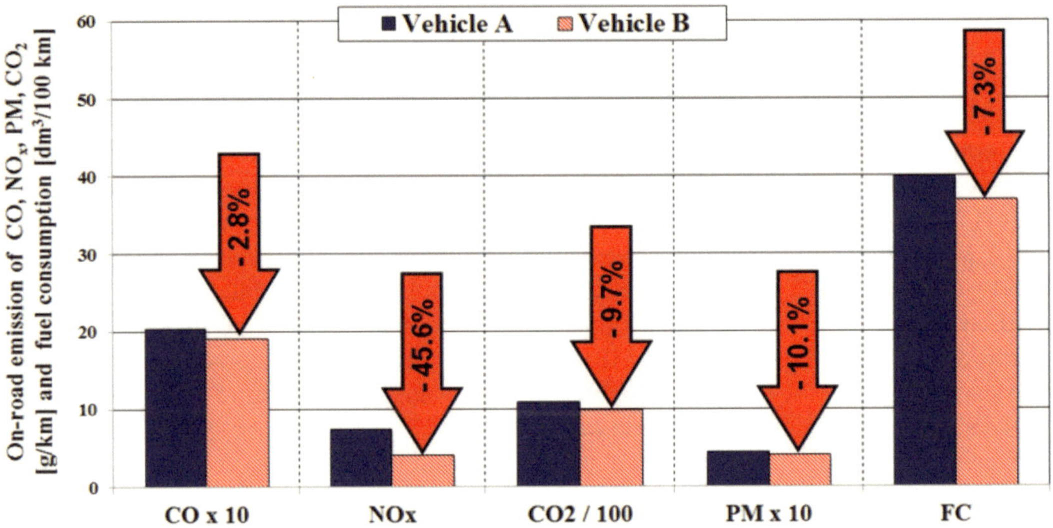

Figure 22. Comparison of the on-road emission of CO, NO_x, CO_2, PM and gas mileage of the tested vehicles.

in the gas mileage (by 2.9 dm^3/100 km). The cost of fuel is currently the main cost of operation of long-haulage trucks. The greatest drop was observed for the on-road emission of NO_x, which mainly resulted from the application of the SCR system in vehicle B.

The reduction of exhaust emissions requires a continuous search for new solutions in terms of both the design of engines/powertrains and the methodology of their testing. A factor stimulating this advancement is the exhaust emissions legislation. The advancement of the exhaust emissions measurement technology has created new possibilities in terms of measurements performed under actual conditions of operation. Supposedly, this particular method will be further developed and will gain in importance. A natural reaction of the legislators and manufacturers should be the recognition of measurements under actual operation as one of the main methods of homologation testing. Relevant works aiming at the introduction of such changes should finish without delay and the resultant legislation should be of global outreach.

The PEMS-based measurements have provided invaluable information regarding the emissions under actual operation of vehicles including their operating parameters. One of the most important observations is the difference in the emissions between the homologation tests and the tests performed under actual operation. The results of the measurements performed on LDV vehicles indicate significant differences, particularly in terms of NO_x and HC. The reasons for that are different parameters in the homologation tests and those under actual operation. The differences for the HDV Euro III and Euro V compliant vehicles are big, particularly in terms of NO_x and PM. They amount to 45 and 10%, respectively. However, referring these results to the Euro III (vehicle A) and Euro V (vehicle B) limits, the differences are smaller—the reduction of the emission of NOx and PM in the Euro V standard compared to the Euro III standard is 60% and over 90%, respectively.

7. Conclusions

The reduction of exhaust emissions requires a continuous search for new solutions in both engine design and methods of engine testing. A main factor stimulating this development is the exhaust emission legislation in which they are in progress. The advancement of exhaust emission measurement techniques provides new possibilities of engine and vehicle testing particularly under actual conditions of operation (RDE). One may suppose that this method will become prevalent and will gain significance. The aim of the legislators and manufacturers should be the acknowledgment of the RDE measurements as one of the main methods of homologation testing works. The aim of introducing such changes should be completed as soon as possible and the enforceability of the implemented legislation should be global.

Abbreviations

CAN	controller area network
CF	conformity factor
CI	compress ignition
CNG	compressed natural gas
CVS	constant volume sample
DOC	diesel oxidation catalyst
DPF	diesel particulate filter
EBS	electronic braking system
EEV	enhanced environmentally friendly vehicle
EFM-HS	exhaust flow meter high speed
EGR	exhaust gas recirculation
ELR	European load response
EPA	environment protection agency
ESC	European stationary cycle
ETC	European transient cycle
FC	fuel consumption
FID	flame ionization detector
GPS	global positioning system
HDV	heavy-duty vehicle
IARC	International Agency for Research of Cancer
LAN	local area network
LDV	light-duty vehicle
LPG	liquefied petroleum gas
MSS	micro soot sensor

NDIR	non-dispersive infrared
NDUV	non-dispersive ultraviolet
NEDC	new European driving cycle
NTE	not-to-exceed
PC	passenger car
PEMS	portable emission measurement system
PM	particulate matter
PN	particle number
RDE	real driving emissions
SI	spark ignition
SCR	selective catalytic reduction
t	time
T	torque
TWC	three way catalyst
V	velocity
WHO	World Health Organization
WHSC	World Harmonized Stationary Cycle
WHTC	World Harmonized Transient Cycle
WLTC	Worldwide harmonized light-duty test cycle
WLTP	Worldwide harmonized light vehicles test Procedures
W	work

Author details

Jerzy Merkisz, Piotr Lijewski*, Paweł Fuć, Łukasz Rymaniak and Andrzej Ziółkowski

*Address all correspondence to: piotr.lijewski@put.poznan.pl

Institute of Combustion Engine and Transport, Poznan University of Technology, Poznan, Poland

References

[1] Press Release No 213. IARC: Diesel Engine Exhaust Carcinogenic. 2012. Online document https://www.iarc.fr/en/media-centre/pr/2012/pdfs/pr213_E.pdf. 2012. [Accessed 12-06-2012]

[2] Attfield M, Schleiff P, Lubin J, et al. The diesel exhaust in miners study: A cohort mortality study with emphasis on lung cancer. Journal of National Cancer Institute. **104**(11): 869-883

[3] Kheifets L. Epidemiologic studies of ELF and their contribution to risk assessment. 24th Annual Meeting of the Bioelectromagnetics Society, Quebec, 23-27 June 2002

[4] Silverman DT, Samanic CM, Lubin JH, et al. The diesel exhaust in miners study: A nested case-control study of lung cancer and diesel exhaust. Journal of National Cancer Institute. 2012;**104**(11):855-868

[5] Environmental Protection Agency. Health Assessment Document for Diesel Engine Exhaust. 2002. Online document: http://ofmpub.epa.gov/eims/eimscomm.getfile?p_download_id=36319. [Accessed: 15-12-2015]

[6] Merkisz J, Pielecha J. Selected Remarks about RDE Test. Combustion Engines, No 3/2016 (166)

[7] May J, Favre C, Bosteels D. Emissions from Euro 3 to Euro 6 Light-duty Vehicles Equipped with a Range of Emissions Control Technologies. London: Association for Emissions Control by Catalyst; 2013

[8] Kousoulidou M, Fontaras G, Ntziachristos L, Bonnel P, Samaras Z, Dilara P. Use of portable emissions measurement system (PEMS) for the development and validation of passenger car conformity factors. Atmospheric Environment. 2013;**64**:329-338

[9] Feist MD, Sharp CA, Spears MW. Determination of PEMS measurement allowances for gaseous emissions regulated under the heavy-duty diesel engine in-use testing program part 1-project overview and PEMS evaluation procedures. SAE International Journal of Fuels and Lubricants. 2009;**2**(1):435-454

[10] Merkisz J, Fuc P, Lijewski P, Ziolkowski A, et al. The analysis of exhaust gas thermal energy recovery through a teg generator in city traffic conditions reproduced on a dynamic engine test bed. Journal of Electronic Materials. 2015;**44**(6):1704-1715

[11] Weiss M, Bonnel P, Hummel R, Provenza A, Manfredi U. On-road emissions of light-duty vehicles in Europe. Environmental Science and Technology. 2011;**45**:8575-8581

[12] Sensors Inc.: Emissions Measurement Solutions. SEMTECH® DS On Board In-Use Emissions Analyzer. Erkrath 2010

[13] AVL, AVL Micro Soot Sensor. Transient High Sensitive Soot Measurement. Graz; 2010

[14] TSI Inc.: Particle Instruments. Model 3090 Engine Exhaust Particle Sizer™ Spectrometer, USA 2005

[15] Sensors Inc.: Emissions Measurement Solutions. SEMTECH ECOSTAR Mobile Test Bench. Erkrath; 2015

[16] Continental develops intelligent technologies for transporting. http://www.continline.com/www/download/automotive_de_en/gengene/contact_services/downloads/commercial_vehicveh/powertrain/common/pow_emission_booklet_pdf_en.pdf. [Accessed: 20-12-2016]

[17] Commission Regulation (EC) No 692/2008 of 18 July 2008 Implementing and Amending Regulation (EC) No 715/2007 of the European Parliament and of the Council on

Type-approval of Motor Vehicles with Respect to Emissions from Light Passenger and Commercial Vehicles (Euro 5 and Euro 6) and on Access to Vehicle Repair and Maintenance Information

[18] Merkisz J, Radzimirski S. New European Union regulations concerning pollutants emission from the motor vehicles. Transport Samochodowy. 2011;**2**:41-70

[19] Regulation (EC) No 715/2007 of the European Parliament and of the Council of 20 June 2007 on Type Approval of Motor Vehicles with Respect to Emissions from Light Passenger and Commercial Vehicles (Euro 5 and Euro 6) and on Access to Vehicle Repair and Maintenance Information

[20] United Nations World Forum for Harmonization of Vehicle Regulations: Proposal for a New Global Technical Regulation on the Worldwide Harmonized Light Vehicles Test Procedure (WLTP). ECE/TRANS/WP.29/2014/27

[21] Tutuianu M, Marotta A, Steven H, Ericsson E, Takahiro H, Noriyiki I, Hajime I. Development of a World-wide Worldwide Harmonized Light Duty Driving Test Cycle (WLTC). Technical Report UN/ECE/WP.29/GRPE/WLTP-IG DHC Subgroup. December 2013

[22] AVL M.O.V.E. A New Solutions for the Upcoming EU6c – Real Driving Emissions (RDE) Legislation. Graz; AVL List GmbH; 2014

[23] Commission Regulation (EU) 646/2016 of 20 April 2016 Amending Regulation (EC) No 692/2008 as Regards Emissions from Light Passenger and Commercial Vehicles (Euro 6)

[24] Engine Vehicle Integration – Emission Requirements & Technology Scuola di Dottorato di Ricerca 2010 – Road Vehicle and Engine Engineering Science. Materiały konferencyjne: Combustion, Bolonia 30.12.2010

[25] Worldwide Emissions Standards: Heavy Duty and Off-Highway Vehicles. Delphi Innovation for the Real World 2015/2016

[26] Directive 2005/55/EC of the European Parliament and of the Council of 28 September 2005 on the Approximation of the Laws of the Member States Relating to the Measures to be taken Against the Emission of Gaseous and Particulate Pollutants from Compression-Ignition Engines for Use in Vehicles, and the Emission of Gaseous Pollutants from Positive-ignition Engines Fuelled with Natural gas or Liquefied Petroleum Gas for Use in Vehicles

[27] United Nations Economic Comission for Europe: Global Technical Regulation No. 4: Test Procedure for Compression-Ignition (C.I.) Engines and Positive Ignition (P.I.) Engines Fuelled with Natural Gas (NG) or Liquefied Petroleum Gas (LPG) with Regard to the Emission of Pollutants. ECE/TRANS/180/Add.4, 2007

[28] Commission Regulation (EU) No 582/2011 of 25 May 2011 Implementing and Amending Regulation (EC) No 595/2009 of the European Parliament and of the Council with Respect to Emissions from Heavy Duty Vehicles (Euro VI) and Amending Annexes I and III to Directive 2007/46/EC of the European Parliament and of the Council

[29] Regulation (EC) No 595/2009 of the European Parliament and of the Council of 18 June 2009 on Type-approval of Motor Vehicles and Engines with Respect to Emissions from Heavy Duty Vehicles (Euro VI) and on Access to Vehicle Repair and Maintenance Information and Amending Regulation (EC) No 715/2007 and Directive 2007/46/EC and Repealing Directives 80/1269/EEC, 2005/55/EC and 2005/78/EC

[30] Merkisz J, Fuc P, Ziolkowski A. Impact of masses load for CO_2, NO_x emission and fuel consumption heavy duty vehicles a total mass exceeding about 12 000 kg. Advances in Science and Technology. 2012; No. 15/2012

Chapter 7

Modern Pneumatic and Combustion Hybrid Engines

Wladyslaw Mitianiec

Additional information is available at the end of the chapter

http://dx.doi.org/10.5772/intechopen.69689

Abstract

This chapter presents the possibility of use of the pneumatic piston engine with two-stroke cycle or four stroke-cycle of the work as an alternative driving source or additional power for the battery regeneration in the electric vehicles. Additionally, such engine can work together with combustion engine as a drive unit in a road vehicle. During city driving, such engine is driven by compressed air. The energy for the engine work is taken from the energy of the air stored at high pressure (about 30 MPa) in bottles or tanks. The chapter presents the thermodynamic theory included in the mathematical model of the engine based on thermodynamic processes (mass and energy balance). On such considerations, the chapter shows the results obtained from specially written computer program for the determination of the most important factors. The results of the calculations are included in the graphs showing the influence of the control parameters (air pressure, injection timing and rotational speed) on the engine working parameters. Certain chapters concern to a hybrid combustion system with an air injection only for a compression ignition engine in order to achieve higher indicated mean pressure and lower fuel consumption.

Keywords: pneumatic piston engines, alternative power source, air injection, efficiency, emission, gas flow

1. Introduction

Nowadays, the environmental regulations of the exhaust emission from internal combustion engines are more rigorous every year. Despite the high progress of the new types of combustion processes such homogeneous charge compression ignition (HCCI), controlled auto-ignition (CAI) [1], ATAC [2] and other or applying of different complicated fuel injection systems, the emission of the combustion products of the hydrocarbon fuel is still high, particularly in lower engine loads. Only small energy of fuel (about 25–45%) depending on the engine type is transformed into mechanical power. The application of alternative energy sources and alternative driving system is need instead of those based on fossil fuels. However, the main environmental problem takes place in big cities with transportation vehicles, where only the fossil fuels are used. Recently, the hybrid systems and fuel cell system are considered for future transportation means. Until now the electricity is produced mostly in many countries by burning fossil fuels. It is connected with production of CO_2 and emission of the toxic components of exhaust gases. The electric vehicles have small possibility to drive a long distance. Up to now, the highest distance for such vehicles reaches maximum 150 km at medium speed and load, but real distance is up to 100 km. For that reason, an additional source power for generating an electric energy or driving source is still required. Many works are concerned on range-extender vehicles with a piston engine driving the electrical generator that charges the batteries. The current from batteries is delivered to the electric engine connected with a driving gearbox that transmits power to the wheels. The combustion engine works only outside the city.

The alternative proposition of power source is to apply the air energy stored in the tank at high pressure. The idea of air-powered engines is known from many years. Already in the nineteenth century, were given concepts of such an engine. In 1847, Mr Parsey invented the air-compressed locomotive and after many years in 1896, the conception of Porter's pneumatic locomotive appeared [3]. The idea of pneumatic engines for transportation was revived again at the end of twentieth century. Many scientific and research works on pneumatic piston engines were carried out across the world in the past few years [4–6].

A car using energy stored in compressed air produced by a compressor has been suggested as an environmentally friendly vehicle in the future by Creutzig et al. [7, 8]. They analysed the thermodynamic efficiency of a compressed air car powered by a pneumatic engine and consider the merits of compressed air versus chemical storage of potential energy. Many proposals of applying the air piston engines were presented by researchers from Asia [9, 10]: for application in transportation. The researchers presented theoretical studies on engines of a typical small-scale passenger car, which are used for the analyses, and the comparison is based on the shaft work, cooling, efficiency and energy density. They found that optimization of the internal-combustion and recycling of the exhaust energy can increase the vehicle's efficiency from an original 15 to 33%, an overall increase of 18%. A hybrid pneumatic system with recirculation of exhaust gases was proposed by Huang et al. [11]. Huang et al [12] carried out a modification of four-stroke engine for operation in two-stroke engine, which was fed with compressed air. Their study presents an experimental investigation on a piston engine driven by compressed air. The compressed air engine was a modified 100 cm^3 internal combustion engine obtained from a motorcycle manufacturer. The experimental and theoretical analysis of a compressed air four-stroke engine was conducted by Chinese researchers, Yu and Cai [13]. The results show that the prototype of such an engine has a good economic performance under low speed and when the supply pressure is 2 MPa. Many works concern to application of the air engine in motorcycles [14], particularly in regions where motorcycle is a main transportation source. A prototype was built with a fuzzy logic speed controller and tested on the real road. Another prototype of motorcycle air engine with a capacity of 100 cm^3 was built by Wang et al. [15]. The motorcycle installed with the compressed air engine can operate at a maximum speed of around 38.2 km/h and a distance up to 5 km equipped with two 9 l bottles filled by air under pressure of 25 MPa.

Currently across the world, there are realized several projects of road vehicles with pneumatic drive, including project developed by Motor Development International (MDI) [16], and sold under the name of the Indian company TATA under the name Tata Air Car Mini Cat [17]. This engine operates in a four-stroke cycle and outside air is drawn into the compression chamber and compressed to 20 bar. At top dead centre (TDC), this air reaches 400°C, and at that point, air from the storage tank is injected into the combustion chamber. The compressed air is stored in carbon fibre tanks at 30 MPa. Recently, the company Peugeot has developed a drive internal combustion engine with a hydraulic system driven by compressed air under the name Peugeot 2008 Hybrid Air [18]. This solution continues to include the concept of the engine powered only with air or the two drive units. The proposed solution is also unique in the world because it includes a combustion engine and air in one drive unit, providing a compact whole drive without having to install a complicated powertrain unit as the current hybrid structures.

2. Principles of work of pneumatic engine

2.1. Why air compressed two-stroke engine?

The energy of the air pressure is delivered to the engine in strictly defined period in order to force the piston in the cylinder of almost standard engine. The work cycle follows only when piston moves down. For that case, the best solution is applying of the two-stroke engine, which performs the real work for every rotation of the crankshaft. The two-stroke engine with port timing is cheaper and simply designed compared to the four-stroke engine of the same capacity. Theoretically, the two-stroke engine gives two times higher power than the four-stroke engine, and a direct fuel injection can fulfil environmental requirements [19]. The energy of the compressed air is converted during the expansion process on the mechanical work. The temperature of air stored in the tank is the same as the ambient temperature, thus the energy depends only on the pressure. The temperature can be increased by heating of the air transferred to the cylinder and thus the energy delivered to the cylinder is higher. However, the thermal losses during opening the exhaust port are also higher. The heat exchange with cylinder walls is smaller than in the classic IC two-stroke engine, because the charge temperature inside the cylinder is low even in TDC.

The pneumatic engine works until the pressure in the tank is high enough to fill the cylinder. The value of torque depends on the air mass delivered from the tank trough valve to the cylinder. One of the most important factors influencing the work of the pneumatic engine is valve timing and a value of the air pressure. The pneumatic engine enables the driving of the vehicle with real-zero emission without any combustion process. The vehicle mobility can be increased by adding an additional heat source in order to deliver higher energy to the cylinder. The pneumatic two-stroke engine together with electric engine will fulfil the future environmental requirements. The experimental set-up of the pneumatic engine has been carried in whole across the world and some vehicles appeared for testing on the road.

2.2. Operation of pneumatic engine

The work performed by the pneumatic engine depends on the pressure difference between higher and lower heat source. The air expansion process is shown in **Figure 1** from pressure p_1 to pressure p_2 with temperatures T_1 and T_2, respectively [20]. The thermodynamic process between point 1 and point 2 is non-isentropic process, and the work l_s has lower value than the isentropic process [21]. In order to obtain the higher power during one work cycle, the higher pressure of the higher heat source (tank) is required. If temperature of the air in the tank has value near the ambient temperature T_1 then temperature of the expanded air T_2 has lower temperature than ambient temperature. But, in a real pneumatic engine, the air injection takes place at the maximum value of compressed air delivered to the cylinder during the intake process. Therefore, the temperature T_1 has higher value than ambient temperature.

The engine is filled only by the air at high pressure when the piston is at TDC. The pneumatic engine can be simply done by modification of the design of the classic two-stroke engine. The engine does not require the inlet port delivering the air to the crankcase. The crankcase has a vent that causes only small compression of the air. The oiling of the bearings and the cylinder surface is ensured by a small oil pump or by oil drop valve in a close cycle. The schematic idea of the pneumatic two-stroke engine and timing of valve and port opening are shown in **Figure 2**. Only one exhaust port is used for the gas exchange in the cylinder. The engine has an injector or pneumatic valve controlled by the electronic unit. The bottle of certain volume contains the air at high pressure. The pressure of stored air in the bottle or tank (about 300 bar) is reduced by a pressure regulator to smaller injection pressure about 20–30 bar. The pressure is controlled by the sensor and the air is delivered by the pipe of small diameter (about 5–8 mm) to the valve. The air volumetric flow rate through the valve is rather high in comparison to the liquid fuel injection. The use of the electromagnetic stem valve requires high voltage and high electric power. For that case, the electromagnetic pneumatic valve used in industry is better solution. The air flow control should enable the high pressure in the cylinder after top dead centre (ATDC), and on the other hand, the opening of the pneumatic valve lasts very short (about 40–60° CA) and due to this reason, the natural frequency of the moving elements

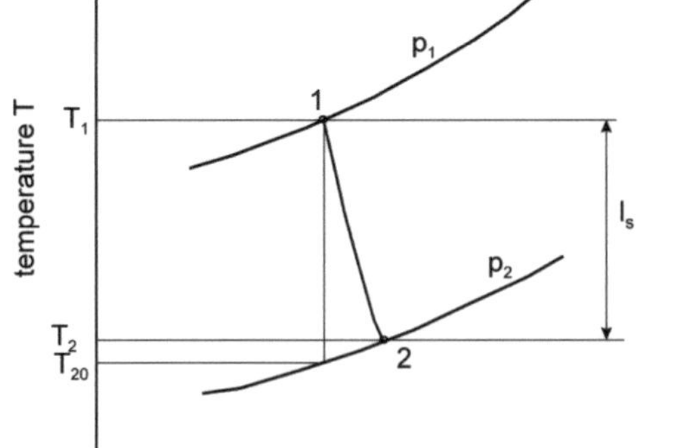

Figure 1. Non-isentropic work during air expansion.

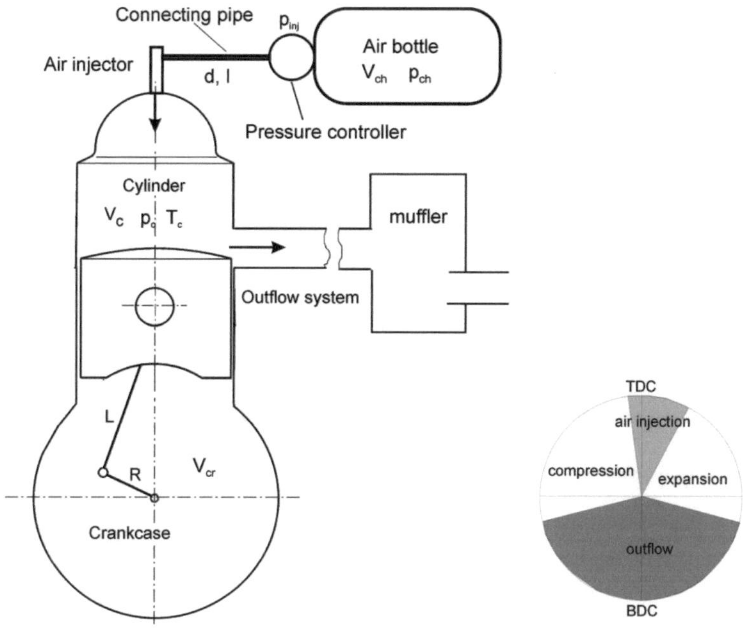

Figure 2. Diagram of two-stroke pneumatic engine and engine timing.

in the valve should be high. The engine is equipped with a muffler for damping the air outflow from the cylinder after opening the exhaust port. Depending on the rotational speed and load, the air injection period begins several degree of CA before TDC. The expansion follows after the air injection and it lasts until the exhaust port opens.

The engine power is controlled only by change of the valve timing. The friction losses, compression stroke, pumping losses in the crankcase and outflow energy decrease the total engine efficiency.

3. Modelling of physical processes in piston pneumatic engine

The mathematical model of the pneumatic engine was carried out to determine the engine performance at different control parameters. Calculation of the air mass delivered to the cylinder by determination of velocity and density of the air in inlet duct in front of the pneumatic valve enables assessment of engine work time at given tank volume and initial pressure. The air was treated as semi-perfect gas, where the specific heat ratio was calculated every time step [22].

3.1. Mass balance

The air thermodynamic parameters in the pipes and ducts were determined at assumption of unsteady gas flow from the three hyperbolic nonlinear partial differential equations: mass, momentum and energy balance. The system of the equations was solved by using the Lax-Harten-Leer scheme [23] based on Godunov's method. The engine parameters were determined on the basis mass and energy balance of the charge in the cylinder. Based on the mass balance law, the increment of the air mass in the cylinder can be expressed by the following equation:

$$dm_c = F_1 \cdot u_1 \cdot \rho_1 \cdot dt + F_{\text{inj}} \cdot u_{\text{inj}} \cdot \rho_{\text{inj}} \cdot dt - F_2 \cdot u_2 \cdot \rho_2 \cdot dt \tag{1}$$

where F_1 = cross section area of the inflow pipe (delivered air from the regulator), F_2 = cross section area of the outflow port, F_{inj} = cross section area of injector nozzle, u = charge velocity in the pipes, ρ = charge density in the pipes, t = time, inj = air injector.

3.2. Conservation of energy

Change of the internal energy U in the time increment dt is determined by the formula:

$$dU = i_1 \cdot dm_1 + i_{\text{inj}} \cdot dm_{\text{inj}} - i_2 \cdot dm_2 + dQ_h - p_c dV \tag{2}$$

where Q_h = heat exchange with walls, i = enthalpy of the air at defined temperature, V = cylinder volume, dm = mass flow rate in the pipes and through the injector, p_c = cylinder pressure.

3.3. Determination of cylinder pressure

After some simplifications and assuming k as specific heats ratio ($k = c_p / c_v$), the energy equation gives the formula of the pressure increment in the cylinder:

$$dp_c = \frac{k-1}{V}\left(dQ_h - kp_c dV + kR(T_1 dm_1 + T_{inj} dm_{inj} - T_2 dm_2)\right) \tag{3}$$

Equation (3) does not contain the component of fuel combustion as in the real combustion engine. All these components depend on time, an increment of the inflow air mass takes place during opening of transfer ports and increment of outflow air mass takes place only during opening of the exhaust port, whereas the air injection lasts very short and begins when piston is near TDC.

3.4. Mass flow rate through injector nozzle

The mass flow rate of the injected air depends on pressure difference between the injector and the cylinder. During calculation, one should check whether the flow is critical or subcritical. For the second case, the mass flow rate is determined from the following equation:

$$\frac{dm_{inj}}{dt} = \frac{\Phi \cdot F_{inj} \cdot p_{inj} \cdot \sqrt{k}}{a_{inj}} \sqrt{\frac{2k}{k-1}\left[\left(\frac{p_c}{p_{inj}}\right)^{\frac{2}{k}} - \left(\frac{p_c}{p_{inj}}\right)^{\frac{k+1}{k}}\right]} \tag{4}$$

where Φ = flow resistance through the injector nozzle, F_{inj} = injector flow area, p_{inj} = pressure of injected air, a_{inj} = sound velocity of injected air ($a_{\text{inj}} = \sqrt{kRT_{\text{inj}}}$), p_c = pressure of air in cylinder and R is an individual constant of air.

In most cases, the air injection takes place at a critical flow, because pressure in the injector is several times higher than in the cylinder. The critical flow occurs when the following condition is fulfilled:

$$\frac{p_c}{p_{inj}} \leq \left(\frac{2}{k+1}\right)^{\frac{k}{k-1}} \tag{5}$$

In such a case, the mass flow rate is calculated as follows:

$$\frac{dm_{inj}}{dt} = \frac{\Phi \cdot F_{inj} \cdot p_{inj} \cdot k}{a_{inj}} \sqrt{\left(\frac{2}{k+1}\right)^{\frac{k+1}{k-1}}} \tag{6}$$

Gaseous constant R of the air in the cylinder does not depend on temperature and k is the specific heat ratio and should be calculated for every considered time step Δt on the basis of the change of temperature in the cylinder and pipes [23] which means that $k = f(T)$.

3.5. Unsteady gas flow in feeding ducts

The mathematical model of gas flow in the pipes and engine ducts takes into account pressure wave motion, which means that flow is unsteady. Non-dimensional velocity $A = u/\hat{a}$ of the air flown into the cylinder is calculated on the thermodynamic equations for isentropic unsteady gas flow through contraction:

a. For sonic flow:

$$\frac{A^2 B^{2c}}{\Psi^2 b^c} - b\left(B^2 + \frac{A^2}{c}\right)^{c+1} = 0 \tag{7}$$

b. For subsonic flow

$$A^2 = c\frac{B^2 - 1}{B^{2c} - \Psi^2}\Psi^2 \tag{8}$$

where $B = \frac{p}{\hat{p}}$, $c = \frac{2}{k-1}$, $b = \frac{2}{k+1}$, $\hat{a}$ = substitute of gas sound speed, $\hat{p}$ = substitute of pressure, Ψ = general flow coefficient.

Gas velocity u is calculated from the given nonlinear equations by solving variable A. The same equations enable the calculation of the air outflow velocity in the pipe near the exhaust port.

3.6. Heat transfer and kinematic dependencies

The amount of the heat transfer to the walls is calculated on the basis of the conductive heat coefficient h_c, area of heat exchange F_h and the temperature difference between gas T_c and walls T_w [24]:

$$dQ_h = -h_c \cdot F_h \cdot (T_c - T_w) \cdot dt \tag{9}$$

After finding cylinder pressure p_c and knowing the charge mass m_c, we can find temperature from the equation of general gas state:

$$T = \frac{p \cdot V}{m \cdot R} \tag{10}$$

Cylinder volume V is determined from the dependency:

$$V = \frac{\varepsilon}{\varepsilon - 1} V_s \tag{11}$$

where ε = compression ratio, V_s = engine piston displacement.

Cylinder piston displacement is determined from kinematics dependencies of the crank-piston system in dependence on crank angle position α:

$$V_s = \frac{\pi \cdot D^2}{8} \cdot s \cdot \left(1 + \frac{\delta}{4} - \cos\alpha - \frac{\delta}{4}\cos 2\alpha\right) \tag{12}$$

where D = piston diameter, s = piston stroke, δ = crank ratio ($\delta = s/(2L)$) and L = connecting rod length. The processes taking place in the cylinder, inlet valve and exhaust pipe are fully described in the literature [21, 22, 25]. The whole model takes into account a wave pressure motion in the pipes and changes of the thermodynamic parameters in each time step (semi-perfect gas). The model enables the calculation of the pressure, temperature, density, air velocity in the inlet and outlet pipes and also the air consumption.

4. Calculation results of two-stroke pneumatic engine

The presented mathematical model was the basis for the development of a computer program in order to simulate the processes taking place in the virtual pneumatic engine. Several works concerning to a piston pneumatic engine were published by Mitianiec and Wiatrak in engine's literature [26, 27]. For this elaboration, the calculations were carried out for different rotational speeds, different filling pressure of the air and valve control parameters.

4.1. Geometrical parameters of engine and boundary conditions

The simulation process considers to the two-stroke engine Robin EC12 with bore D = 75 mm, stroke S = 55 mm and compression ratio 6.5 and opening of exhaust port at 106° CA ATDC, which was fully tested in the standard version. Initial pressure in the tank was assumed as 300 bar. The pipe connecting the reducer and the valve with length 80 mm amounted had 8 mm of diameter. The valve lift during opening was assumed as sinusoidal. The air temperature in the tank was near ambient temperature and amounted 300 K. The higher filling pressure also causes higher cylinder pressure, as is shown in **Figure 3**, at a rotational speed of 2400 rpm.

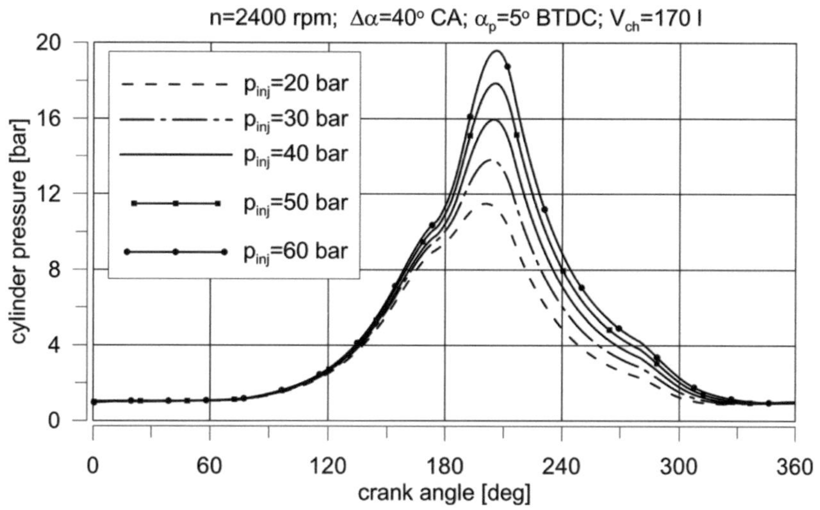

Figure 3. Cylinder pressure in a function of crank angle at different air injection pressure.

The filling pressure of 60 bar causes the pressure increase in the cylinder to maximum value 20 bar. The calculations were carried out at a valve opening 5° CA before top dead centre (BTDC) and duration 40° CA.

4.2. Thermodynamic parameters of pneumatic two-stroke engine

The higher indicated mean effective pressure (imep) value is obtained at lower compression pressure, which takes place at lower compression ratio. The air temperature inside the cylinder depends also on the filling pressure. Higher filling pressure causes higher temperature of the air in the cylinder. Variation of the cylinder temperature is presented in **Figure 4**. The calculations were performed at the same control parameters as for pressure calculations. One can observe very low temperature at the end of the expansion process. At low filling pressure, for example 20 bar, the cylinder temperature decreases below 200 K.

At higher injection pressure, temperature in the cylinder decreases and at value of 60 bar, the maximum temperature at TDC is below 400 K. This situation causes the transfer of heat from the walls to the charge in the cylinder. The characteristic of engine effective power has quite different variation than characteristic of the classic two-stroke engine (**Figure 5**). The engine has bigger power at low rotational speed at the same valve timing. The characteristic was obtained for air injection pressure of 25 bar. In the pneumatic two-stroke engine, the highest value of brake mean effective pressure (bmep) takes place at lowest rotational speeds. This phenomenon is like as in electrical engines. Higher bmep value at higher rotational speeds can be assured by higher filling pressure, which causes a bigger air dose injected by the valve to the cylinder. Another way of increasing of bmep is increasing the duration of valve opening at the same filling pressure. The increase of the air injection pressure causes almost linear increase of bmep (**Figure 6**), but this causes increase of specific air consumption (SAC) value. The

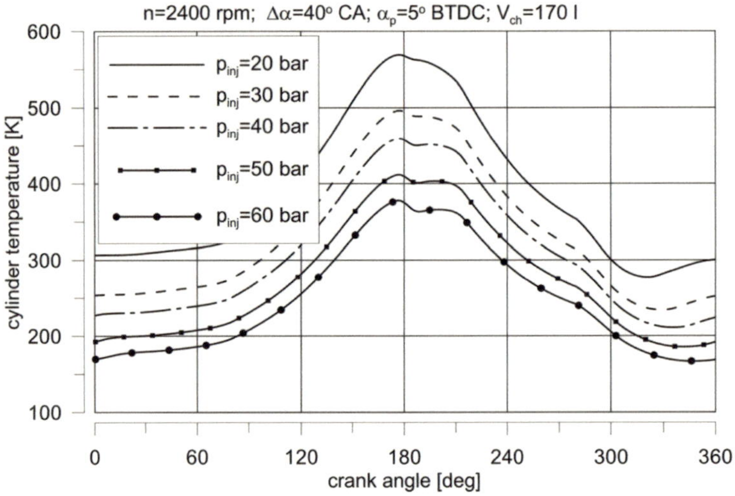

Figure 4. Cylinder temperature in a function of crank angle at different air injection pressure.

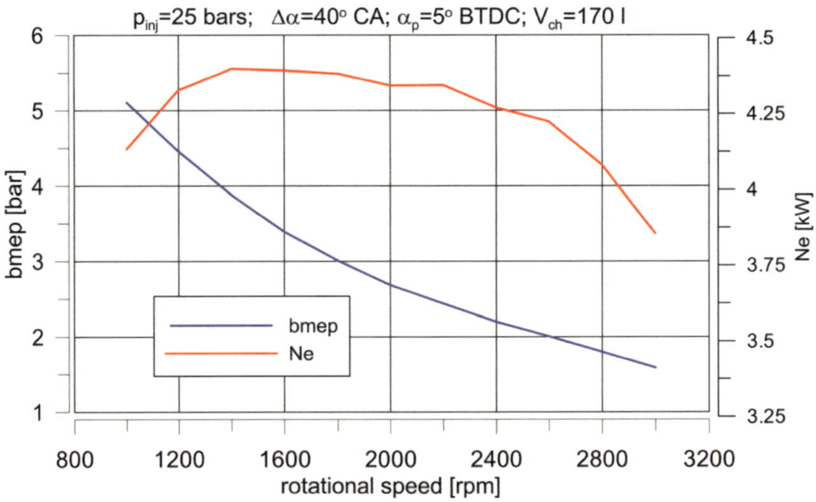

Figure 5. Effective power and torque in a function of rotational speed.

calculations were carried out for valve opening 5° CA BTDC, n = 2400 rpm, wide opening throttle (WOT) and the opening duration 40° CA.

The change of the engine torque requires an automatic control of the filling pressure in the reducer. The same graph shows variation of SAC, which depends linearly on the injection pressure. Lower specific SAC values occur at lower pressure of air injection. Thus very important is reduction of air pressure from high value in the bottle to lower value in the injector.

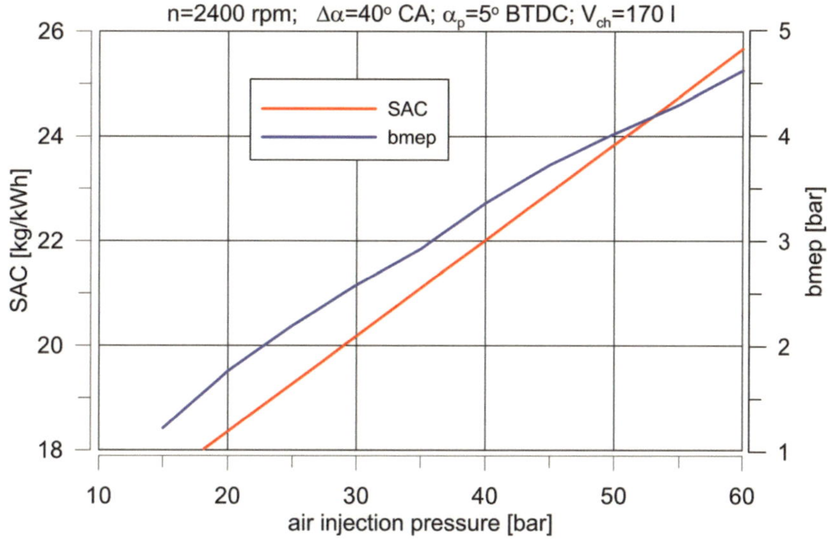

Figure 6. Torque and specific air consumption in a function of injection pressure.

4.3. Air consumption in pneumatic two-stroke engine

As internal combustion engines, the efficiency of the pneumatic engine can be determined by an amount of air mass needed for producing power unit. During valve opening, the air mass delivered to the cylinder was calculated as a sum of partial masses at every time step. Variations of SAC and air mass per cycle (AMPC) as a function of engine rotational speeds are presented in **Figure 7** at an injection pressure of 25 bar. With increasing of rotational speed at the same angle of opening of the injector (shorter time), one can observe a decrease of air mass consumption per cycle, but the amount of cycles increases in the same period.

For higher specific air consumption, the total efficiency is lower at higher rotational speed. This indicates to use the pneumatic engine at lower rotational speeds. The simulation showed the dependence of emptying of the tank on the air injection pressure. Variation of time emptying in a function of the filling pressure is shown in **Figure 8** at a rotational speed of 2400 rpm. Emptying time of the tank at air injection pressure of 20 bars for 55 minutes for the tank with volume 170 l and emptying time of tank with another volume will be almost proportional to that volume.

The simulation shows a small change of the emptying time at medium rotational speeds and constant air injection pressure. It is caused by influence of non-steady gas flow on the ducts of the two-stroke engine. This time is almost proportional to the tank volume. **Figure 9** shows the variation of emptying time as a function of the engine rotational speed at air injection pressure of 25 bar of a tank with volume 100 l at full engine load (WOT). Emptying time does not change rapidly for considered volume of the bottle.

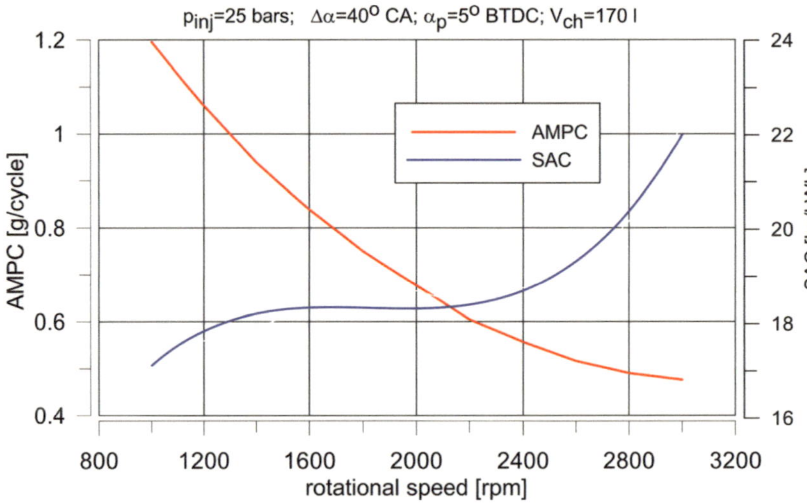

Figure 7. Consumption of air mass per cycle and specific air consumption in a function of rotational speed at air injection pressure 25 bar.

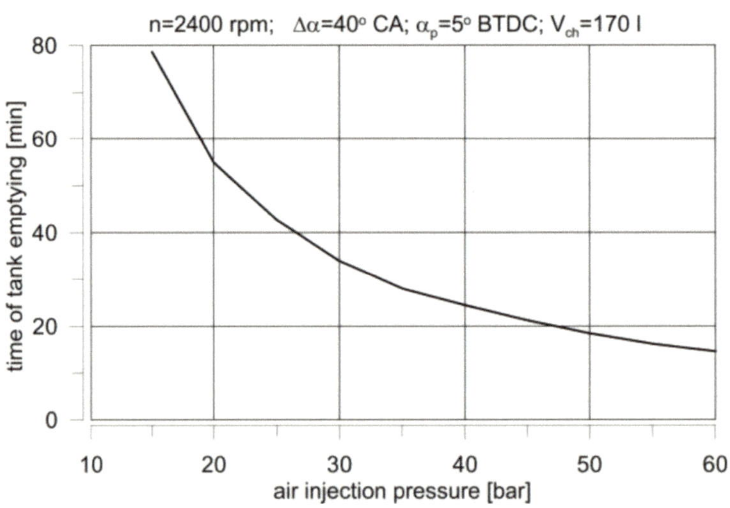

Figure 8. Emptying time of air for a tank with volume 170 l at n = 2400 rpm and different injection pressure and full load.

The pneumatic valve controlled by the electronic unit must enable an adequate air mass flow rate in a short time. The dose of air per cycle is one of the most important parameters needed for the design of the pneumatic valve.

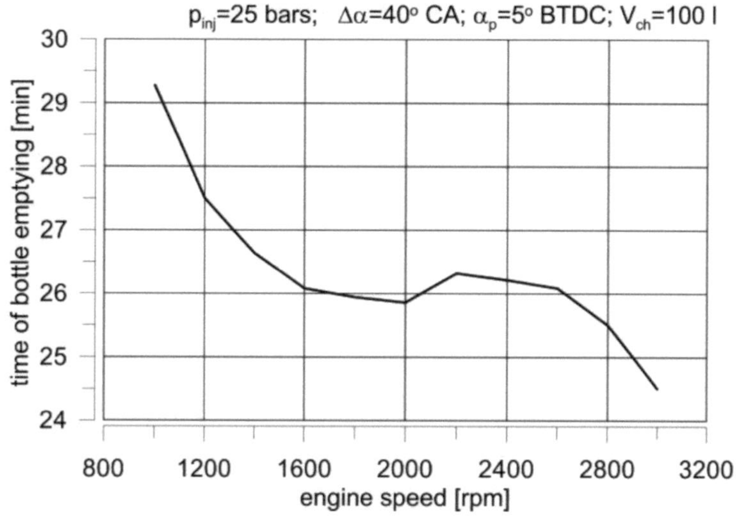

Figure 9. Emptying time of a tank with volume 100 l in a function of engine speed at air pressure 25 bar and full load (WOT).

5. Assessment of two-stroke pneumatic engine

The numerical analysis of the work of the pneumatic two-stroke engine based on the mathematical model and results from the simulation was carried out by theoretical considerations and results obtained from computer program. Most of previously done research works presented in introduction and this work was concentrated on air two-stroke engines. On the basis of the simulation results, the following remarks can be drawn:

1. The two-stroke engine with air injection indicates smoother work than the four-stroke engine for the same air injection pressure and enables better utilization of the compressed air than the rotor engine. Every rotation of crankshaft produces power and such engine only wasted the air to the exhaust port during the scavenge process, not during air injection. For the same rotational speed and pressure of air injection below 150 bar, the four-stroke engine indicates lower bmep and also lower SAC than the two-stroke engine. This air comes from the process of normal filling of the cylinder.

2. Every pneumatic engine indicates higher value of bmep at lower rotational speeds and enables better starting of vehicle. Good characteristic of the two-stroke engine depends on very short crank angle injection with value near 40° CA and start of air injection about 5° CA.

3. The time of engine work depends on the air filling pressure and tank volume. It should be emphasized short time operation of the pneumatic engine under heavy loads. It would add the tanks of greater volume or more resistant (tanks made from composites) to greater pressure to about 700 bar.

4. The increase of the engine torque can be assured by the increase of the air injection pressure but it causes higher specific air consumption.
5. Low temperature at the end of expansion process should not cause a lubrication problem, because the mean temperature of the charge is near the ambient temperature. The wall temperature in every two-stroke engine is stable for steady load. In such case, the engine does not require any cooling system. Lubrication in such engine should be carried out by dosing of lubrication oil to the inlet pipe by a special needle valve with possible regulation of mass flow rate.

The design of the pneumatic engine is based on the classic two-stroke engine and it needs small changes in order to mount the air feeding system with electronic control unit (ECU) system.

6. Practical solutions of pneumatic piston engines

The pneumatic two-stroke piston engine is a simple solution of a real non-conventional driving system, which does not produce toxic gaseous components and does not burn any fuel. However, for filling the bottle with the air under high pressure, a mechanical energy is required for driving of a compressor. It is mostly a piston compressor allowing for obtaining high pressure driven by an electric motor, and energy is obtained from power station (combustion of coal, biomass, nuclear energy, gasification of coal).

6.1. Control valves of air inflow

The important element of every pneumatic engine is the pneumatic valve, which delivers the air in defined time to the cylinder. The air is injected by the valve to the cylinder, where the air charge is compressed due to the movement of the piston towards TDC. The greater outlet area of the injector is required for a large mass flow rate of air. For such case, the poppet valve in the injector sometimes is used. The proposal of the injector with poppet valve is shown in **Figure 10**, where the movement of the valve is controlled by a cam mechanism. Because of high pressure from the inlet side, the moving part of the valve has two parts: poppet and cylindrical parts with the same diameter. This arrangement allows for maintenance of the valve in the closed condition without additional force and closing of the poppet valve required a small force of the return spring. The piston of the valve has a labyrinth sealing and also sealing between the valve stem and the body. The whole controlled mechanism is like as in the timing mechanism in the four-stroke engine. The cam system gives a constant angle of opening of the valve (in CA deg) in relation to crankshaft rotation. Better solution of controlled motion of poppet valve in the air injector is applying of an electronic unit by using an induction coil where electromagnetic forces enable the movement of valve stem. The simpler but practical solution of the air injector controlled by solenoid unit is shown in **Figure 11a**. The valve stem should have a limiter of movement. Electromagnetic force for opening of valve lasts very short and is independent of the movement and position of the crankshaft. The poppet valve can be opened and closed at any time. Another proposed solution is utilization of the standard

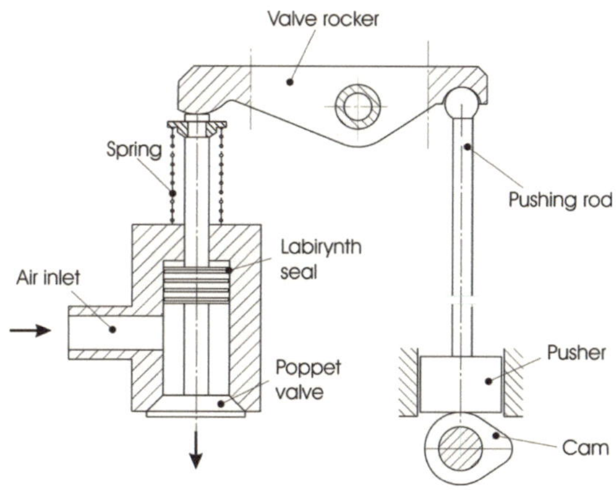

Figure 10. Proposal of air dosing valve driven by a cam mechanism.

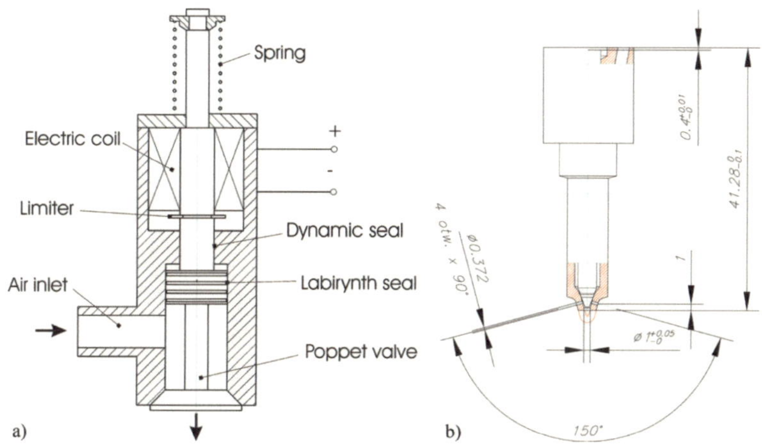

Figure 11. Electromagnetic valves: (a) air poppet valve controlled by ECU and electric coil and (b) utilizing of standard sprayer of CI engine.

sprayer of diesel oil with higher nozzle diameter and by cutting of the dispenser cap, but with contact of the needle spray with the body (**Figure 11b**).

Such design of an injector tip revamped is presented in **Figure 11b** and it was applied by Wiatrak [30] in his pneumatic engine in moped. This configuration of the air injector does not require any modification of ECU and allows control of air injection in any way. Limitation of applying such injector tip is only the required air mass flow rate.

6.2. Two-stroke Indian pneumatic engine

Some solutions of pneumatic engines can be found in the literature and widely in Internet. One of the promising solutions was given by Kumar et al. [28]. They tested a two-stroke engine equipped with storage cylinder, pressure regulator, air filter and lubricator, modified flywheel

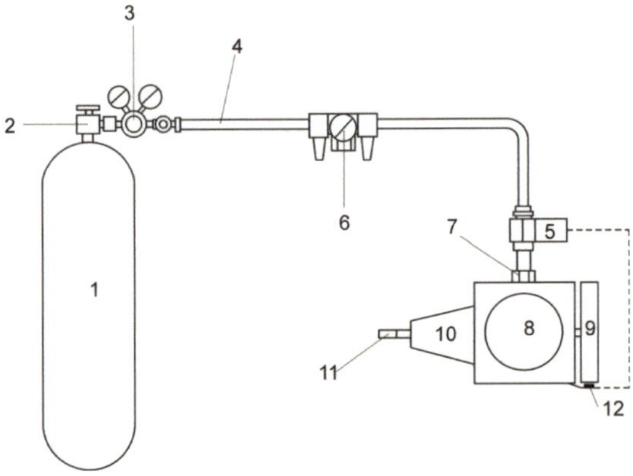

Figure 12. Scheme of compressed air engine: 1, storage cylinder; 2, stop valve; 3, pressure regulator; 4, hose; 5, solenoid valve; 6, air filter and lubricator; 7, adapter nipple; 8, two-stroke SI engine; 9, flywheel; 10, gearbox; 11, transmission shaft; 12, magnetic sensor [28].

and solenoid valve working with 24 V DC with a maximum pressure of 10 bar. A schematic diagram of compressed air engine worked out by Indian scientists is shown in **Figure 12**.

6.3. Small power air engine

Another solution was presented by Chinese scientists, represented by Xu et al. [29], in their work concerning an adaptation of four-stroke engine working on compressed air. They have developed a mathematical model of filling the cylinder and control model of the engine. The virtual model of the pneumatic engine is shown in **Figure 13**. Their work concerns mainly to theoretical analysis of working performance also in dynamic loads by using SIMULINK. The

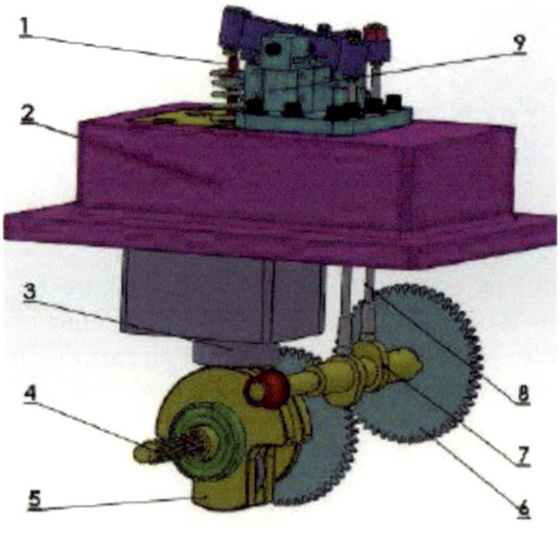

Figure 13. Physical model of APE: 1, balanced valves; 2, cylinder cover; 3, cylinder; 4, crankshaft; 5, crank piston mechanism; 6, timing gears; 7, camshaft; 8, cam follower; 9, tunable rocker mechanism [29].

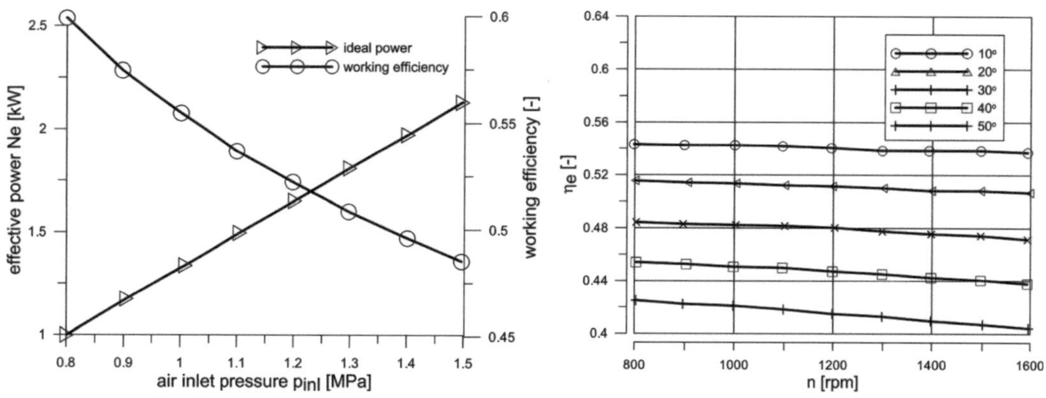

Figure 14. Engine performance: (a) ideal power and working efficiency at different air inlet pressure and (b) working efficiency of APE at different rotational speeds [29].

Chinese researchers conducted numerous analyses of a small engine to find an optimal work efficiency, high torque and an optimal control parameters of injection of the air. **Figure 14a** presents an ideal power and engine efficiency as a function of pressure of the injected air. The engine power decreases with increasing of the air pressure, but engine efficiency linearly increases. It should pay attention to high efficiency above 56% at a pressure inlet of 1.5 MPa.

Engine efficiency depends mainly on timing of the injection valve. **Figure 14b** presents the total efficiency of air powered engines (APE) for different duration of the air injection from 10° to 50° CA at a starting point of injection 5° CA BTDC. Engine efficiency slightly decreases with increasing of rotational speed and for longer opening of the air injector, the engine efficiency decreases from 54 to 42% at an engine rotational speed of 800 rpm.

The Chinese researcher found that 'the virtual prototype of the APE can make the simulation more precise and reduce the cost of the design. This research can provide theoretical supports to the new APE prototype's design and optimization' [16].

7. Pneumatic and combustion hybrid engine

7.1. Conception of vehicle with pneumatic and combustion engine

The proposal concerns to a certain hybrid combustion system in internal combustion engines both compression and spark ignition (SI) in order to achieve higher indicated mean pressure and lower fuel consumption. The solution is combination of two fuelling systems: the first direct fuel injection and the second high pressure air injection. The dosing of both fluids is shifted in CA one relative to second. The additional air helps in the charge mixing, increasing of charge turbulence and causes a quicker combustion process by additional oxygen in the regions, where local excess air coefficient is small (below ignition boundary). Besides the fuel dose, the additional air increases the mass of charge in the cylinder causing a significant increment of pressure. This elaboration concerns only to applying of an air injection in compression ignition (CI) engines. The influence of an additional air dose on compression ignition engines can help to break the fuel jet with possibility to burn the droplets in the kernel of fuel jet. In this way, CI engine can reduce the amount of emitted soot and nanoparticles. The presented solution of combustion and pneumatic engine is based on the patent applications made by the Wiatrak and Mitianiec [30, 31]. The simple diagram of the solution is presented in **Figure 15** for CI engine, but the same solution can be applied for SI engine and in this case instead of the diesel oil injector a spark plug will be located.

A high-pressure bottle contains the air under pressure below 500 bar, and the air is supplied to the pneumatic injector through the safe valve, pressure controller, which reduces high pressure according to the engine load. The electrical signal from ECU controls both the air pressure and time of opening and closing of the pneumatic valve.

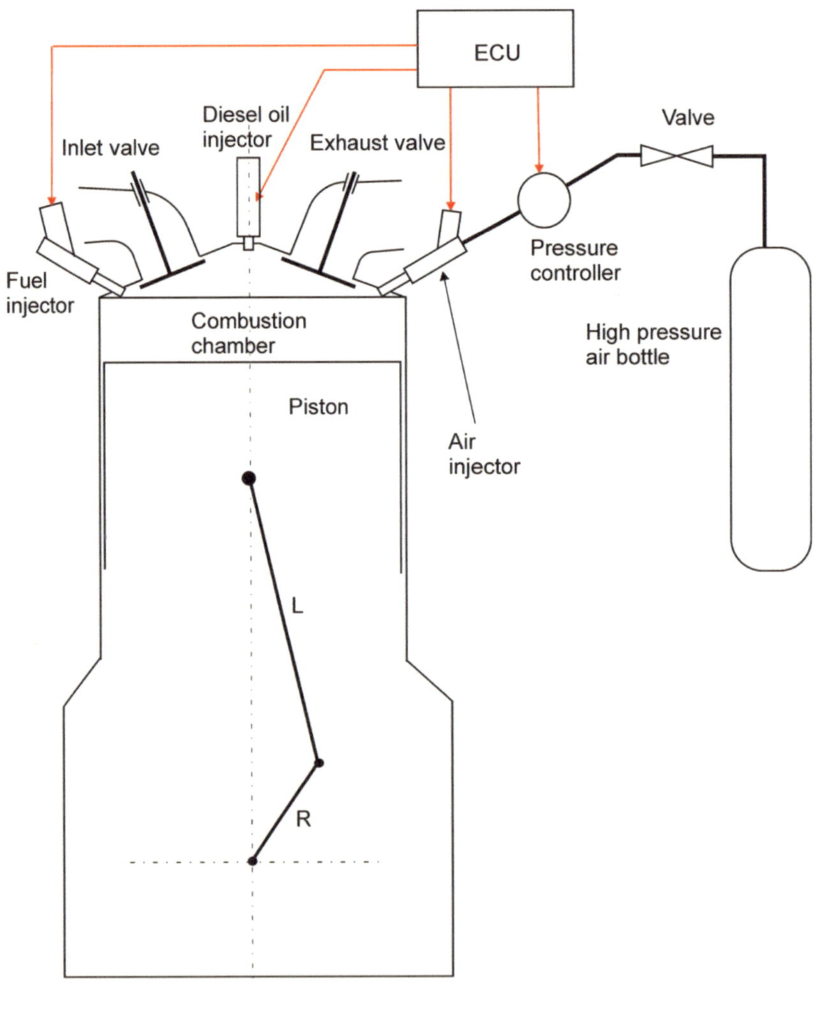

Figure 15. Diagram of combustion and pneumatic CI engine [31].

The electrical signal from ECU controls both the air pressure and time of opening and closing of the pneumatic valve. The electronic control of fuel injector and air injector enables to obtain optimal engine parameters by a small amount of fuel and air. The car with combustion and pneumatic engine can be driven by the piston engine working in four modes as follows:

1. Pneumatic mode (air injection at high pressure to the cylinder) without combustion.
2. Combustion (CI or SI) engine (standard mode).
3. Combustion and full pneumatic mode for temporary high power during acceleration, climbing and high velocity.
4. Combustion with micro-dose of injected air for small increase of car load.

The combustion and pneumatic engine depending on working modes enables the following factors:

- Increasing of engine torque (power).
- Decreasing of specific fuel consumption.
- Driving of the vehicle with real-zero emission as a result of lower temperature (decreasing of NOx emission) and full fuel combustion by adding more air, which enables CO absence.
- Decreasing of soot emission in diesel engines as a result of higher concentration of oxygen in the core of fuel jets (by optimal direction of injected air).
- Decreasing of cooling heat from engine to the cooling system.

The driving resistance power depends on rolling forces and air resistance forces. The resistance force was measured by the Netherlands automotive research institution (TNO) [32] for car with mass 1700 kg and the following formula was given:

$$F = 114.22 + 0.3861 \cdot v + 0.0281 \cdot v^2 \text{ [N]} \tag{13}$$

where v is velocity of car on flat road in km/h. The car with mass 1250 kg driving with velocity 50 km/h on the flat road requires only 2.83 kW of power. Maximum power of engine which was tested according to New European Driving Cycle (NEDC) (t = 400 s) amounts 19 kW and engine work during the NEDC test reaches value 0.314 kWh. Another formula for calculation of driving forces of passenger cars in NEDC was given by the Austrian automotive research company founded by Helmut List (AVL), on the basis of their measurements:

$$F = 102 + 6376.5 \cdot \sin\alpha + 0.02592 \cdot v^2 \text{ [N]} \tag{14}$$

where α is an angle of inclination of the road and v is velocity of the car in km/h.

7.2. Pneumatic CI engine

For simulation of combustion and air injection process in internal combustion engine (ICE), a diesel engine being in production was chosen and some results of simulations are presented below. Simulation was carried out only on one-cylinder compression ignition (CI) four-stroke engine with a capacity of 450 cc. The results of calculations were obtained from the computer program by using 0-D thermodynamic model of engine work cycle with unsteady gas flow in engine pipes.

Technical data of engine:

Bore/stroke = 82/85 mm

Length of connected rod = 130 mm

Compression ratio = 16

Number of valves = 4

Inlet valve timing = 20° BTDC/35° ABDC

Exhaust valve timing = 56° BBDC/20° ATDC

Parameters of air injection:

Pressure = 350 bar

Timing of injector = TDC/35° ATDC

Flow area of valve exit = 5 mm^2.

The diagram shown in **Figure 16** presents variation of mean effective pressure and specific air consumption, which was fuelled only by the air at high injection pressure. Variation of these parameters is the same as for the two-stroke pneumatic engine. When the engine is supplied only with air, bmep rapidly decreases with growing rotational speed (lower value of torque).

The air mass consumption per one cycle decreases with rotational speed at the same air injection parameters, however, the specific air consumption increases with rotational speed.

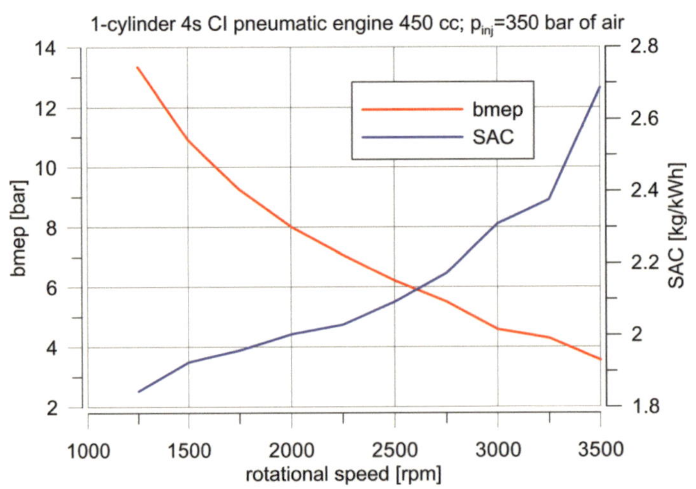

Figure 16. Mean effective pressure and specific air consumption of one-cylinder CI four-stroke engine fuelled only by air injection at injection pressure 350 bar.

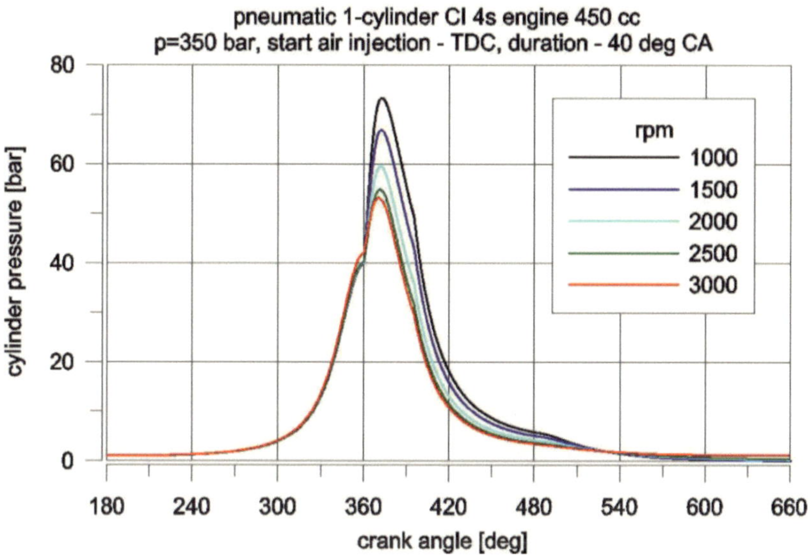

Figure 17. Cylinder pressure traces in one-cylinder CI four-stroke engine fuelled only by air injection at pressure 350 bar for different rotational speeds.

At higher rotational speeds, the pneumatic engine has lower efficiency at the same air injection parameters. With increasing of engine rotational speed, the cylinder pressure decreases, because the time of air injection is shorter for the same duration of crank angle of the air valve opening.

Figure 17 shows traces of pressure in one cylinder for considered CI pneumatic engine at different rotational speeds. The higher maximum pressure in the diesel engine supplied only with air takes place at lower rotational speeds. With the increase of engine speed, the maximum of air pressure in the cylinder still decreases. Calculations were carried out for CI engine at start of air injection at TDC, duration of injection 40° CA and air injection pressure 350 bar. The diesel pneumatic engine indicates higher specific air consumption at higher rotational speed, where the engine power has lower value. These parameters show that the pneumatic diesel engine indicates better working parameters at lower rotational speeds.

7.3. Combustion engine with additional dose of injected air

Simulation of CI engine with the same geometrical parameters as in the first option with additional air injection (350 bar) was carried out for different rotational speed. Variation of cylinder pressure is shown in **Figure 18** for the CI engine fuelled only by diesel oil and for CI engine (the same air excess ratio $\lambda = 1.5$) with additional air injection. It is seen higher pressure in the cylinder during expansion stroke for the combustion and pneumatic mode than for CI engines only. Start of air injection was constant for all presented rotational speeds: opening 25° CA ATDC and duration 40° CA.

Figure 19 presents the variation of engine torque for both cases as a function of rotational speed. The big difference of engine torque and also engine power is observed particularly at

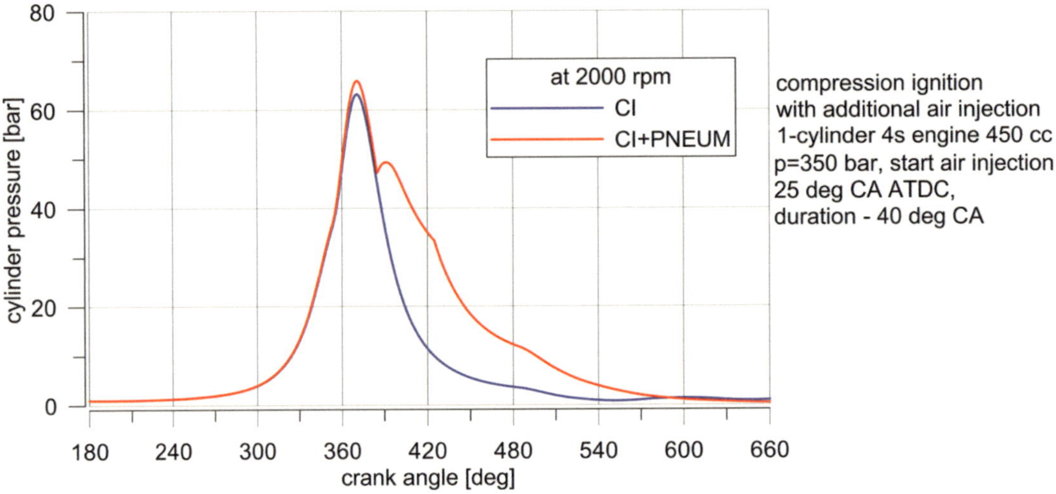

Figure 18. Comparison of cylinder pressure in the four-stroke CI engine and CI engine with additional air injection at 2000 rpm.

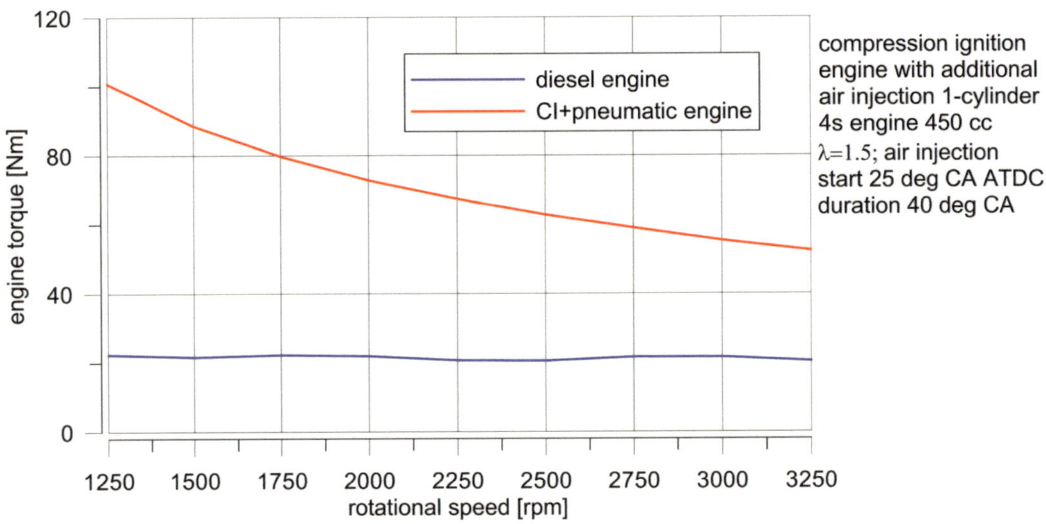

Figure 19. Comparison of engine torque in the four-stroke CI engine and CI engine with additional air injection at different rotational speeds.

lower rotational speed, because in the pneumatic engine at the same duration of opening, the air injector more air is delivered to the cylinder in lower rotational speed because of longer time of opening of the air injector.

By adding the pressured air into small volume of the combustion chamber (almost at TDC), the engine power rapidly grows and for that reason the specific fuel consumption (diesel oil) decreases considerably, as shown in **Figure 20** for engine with additional dose of the air. This

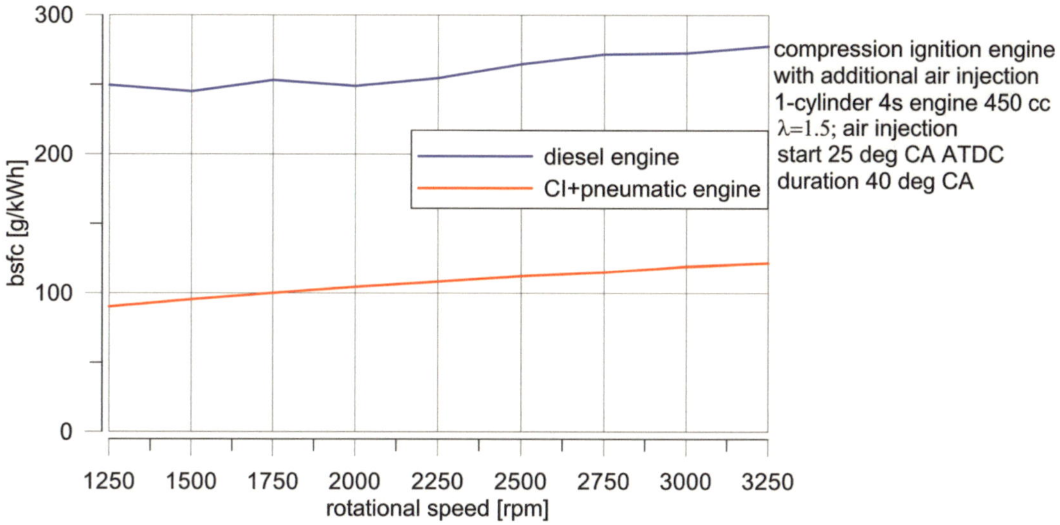

Figure 20. Comparison of specific fuel consumption in the four-stroke CI engine and CI engine with additional air injection at different rotational speeds.

mode of engine work is needed for higher acceleration and load of the car and enables the retrieving of very high power density with dose of the air for only short time of operation. At higher rotational speeds, only small dose of the air is required.

This mode of engine work is needed for higher acceleration and load of the car and enables the retrieving of very high power density with dose of the air for only short time of operation. At higher rotational speeds, only small dose of the air is required.

7.4. ICE with micro-dose of air

One of the possibilities of the combustion and pneumatic engine is working with a micro-dose of the pressurized air during normal compression ignition engine operation (combustion mode). The micro-dose of air is required, for example, by reducing CO emission or reducing combustion temperature, which influences on decreasing of NOx emission. However, this small dose of air increases significantly the engine power. This mode can be fulfilled by changing of duration of air injection. **Figure 21** shows the variation of engine effective power with micro-dose of the air in comparison to the standard engine as a function of duration of the air valve opening. The specific fuel consumption for such working mode is considerably reduced (**Figure 22**). The presented results were calculated for 2500 rpm and at an air pressure of 350 bar. The value of brake specific fuel consumption (bsfc) decreases rapidly as a result of a longer opening of air valve and delivering more air to the cylinder.

7.5. Strategy of control of air injection

Change of pressure trace in the cylinder in the combustion and pneumatic engine depends on timing of the air valve. Variation of cylinder pressure is shown in **Figure 23** for the same pressure of injected air (150 bar) but at different valve opening and at the same duration of

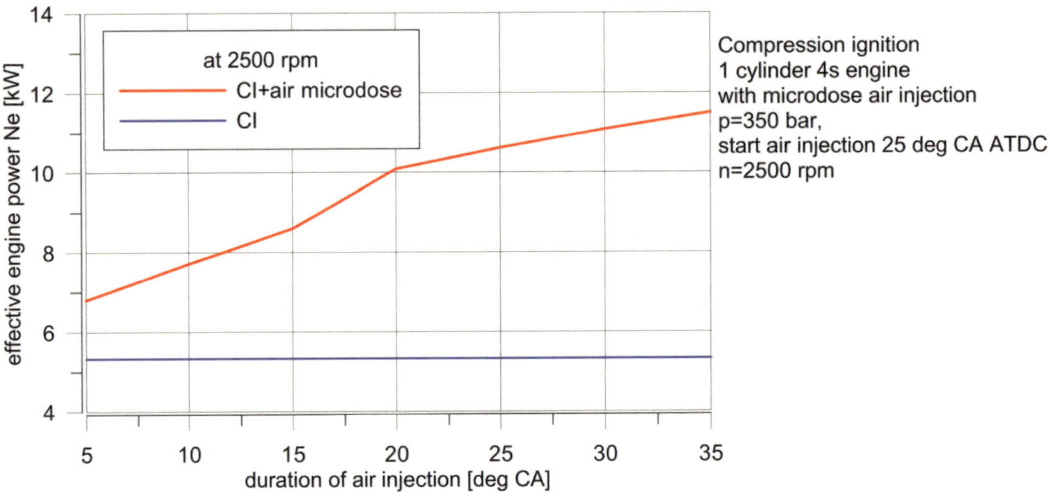

Figure 21. Influence of duration of air injection on engine effective power at 2500 rpm.

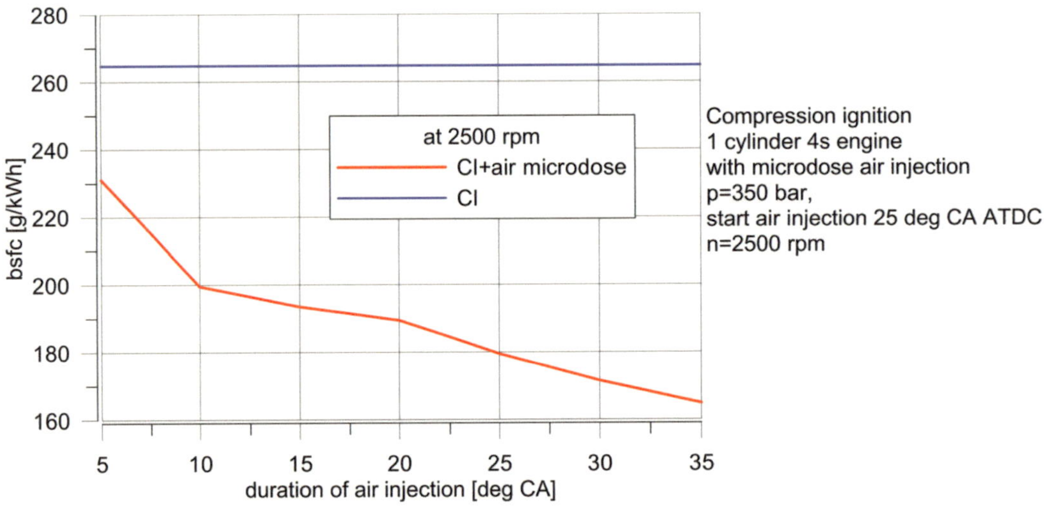

Figure 22. Influence of duration of air injection on specific fuel consumption at 2500 rpm.

valve opening. By earlier opening of the air valve, one obtains a higher mean indicated pressure. The control of engine work can be realized by changing of timing of air injector opening for constant pressure. At high inlet air pressure, the flow is critical (sonic) and air velocity is constant and equalled to the local sound speed. The mass dose of the air can be controlled also by changing of the pressure, however, such regulation is not suitable for this application. The air is stored in the pressurized tank or bottles, which can be located in different places in the car. Simulation carried out in GT-Power program indicates lower emission of carbon monoxide in SI engine with air-added injection than in standard SI engine, because of higher concentration of oxygen in the combustion chamber (**Figure 24**).

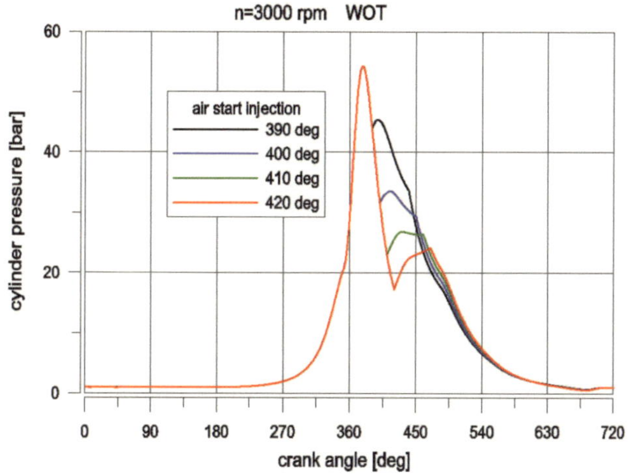

Figure 23. Pressure variation in a cylinder for different angles of air injection start at 3000 rpm in the four-stroke CI engine (combustion + air injection).

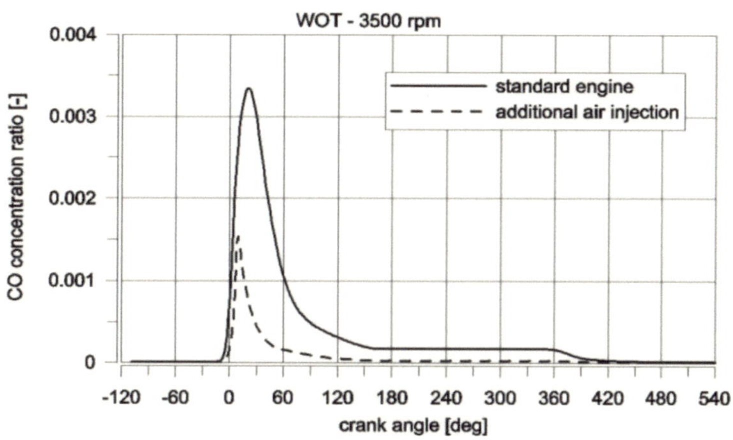

Figure 24. Comparison of cylinder CO mass concentration at 3500 rpm for standard four-stroke CI and hybrid engine at full load.

The pneumatic system also enables the reducing of soot emission in diesel engines by earlier air injection in the region of the fuel jet core, where the amount of oxygen is not enough for complete burning and thus the soot is formed. Simulation of work performance was also carried out for SI engine with a compression ratio $\varepsilon = 10$ for the same geometrical data as for the compression ignition combustion and pneumatic engine. The obtained results have the same tendency of increasing power and decreasing of specific fuel consumption. The main task of the proposed system is applying of vehicles without exhaust pollution in the cities ('zero emission') and enabling a higher engine performance at different load of the car.

8. Car pneumatic system

8.1. Proposal of combustion and pneumatic hybrid vehicle

The proposal of the hybrid car driven by combustion and pneumatic engine is shown in **Figure 25**. During normal driving on the highway or high speed road (outside of the city), the diesel or gasoline engine drives the piston air compressor, which loads the air to the tank. The work done for the air compression is recovered during driving in the city or car acceleration. The system has possibility of filling the air tank during stop in the garage or on special filling stations. The air pressure is controlled by a regulator. Each cylinder or only chosen cylinders are equipped with the air injectors, which are controlled by ECU (electronic control unit). The air injection is applied only in the cities in special regions for the 'zero emission' mode on the distance maximum 10 km or higher and for sudden change of engine load. Some energy for driving of the compressor can be obtained from braking energy (recovering energy) during deceleration of the car. It requires some changing in the electric control system by applying of an electric motor for the piston compressor drive.

8.2. Required amount of air during urban drive

For driving in the city with mean velocity v = 50 km/h, the car with mass m = 1250 kg requires power P about 4 kW. Such value of power can be obtained in the CI pneumatic engine at rotational speed above n = 3500 rpm with air consumption in one cylinder m_{cycle} = 0.12 g/cycle.

The amount of air injection pulses during road distance l = 10 km:

$$i = \frac{60 \cdot l \cdot n}{2 \cdot v} = \frac{60 \cdot 10 \cdot 3500}{2 \cdot 50} = 21,000 \text{ cycles} \tag{15}$$

The required mass of air for pneumatic driving:

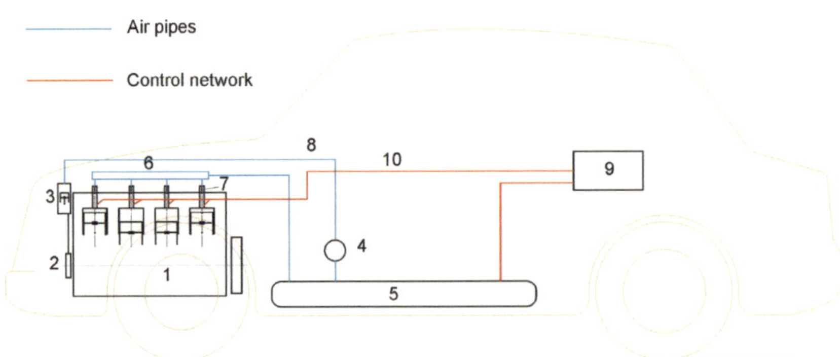

Figure 25. Scheme of pneumatic system of combustion engine in the car. 1, engine; 2, clutch of compressor drive; 3, piston compressor of high pressure; 4, pressure regulator; 5, high pressure air tank; 6, air common rail; 7, air electromagnetic injector or valve; 8, air high pressure line; 9, ECU; 10, electronic control network.

$$m_{air} = i \cdot m_{cycle} = 21,000 \times 0.12 = 2520 \text{ g} = 2.520 \text{ kg} \tag{16}$$

This mass value corresponds to volume of the tank filled with air at pressure 350 bar and temperature 300 K:

$$V = \frac{m_{air}}{\rho_{air}} = \frac{m_{air}RT}{p} = \frac{2.520 \cdot 287 \cdot 300}{350 \cdot 10^5} = 0.0062 \text{ m}^3 = 6.2 \text{ l} \tag{17}$$

The higher volume is required because during consumption of air by the engine, the air pressure in the tank decreases. The calculations carried out by the authors indicated that a required volume of the tank not greater than 30 l. The required bottle volume of air with pressure 350 bar for the following road distance at a constant velocity of 50 km/h:

1. 5 km—3.1 l
2. 10 km—6.2 l
3. 15 km—9.3 l
4. 20 km—12.4 l

For driving of the car according to drive test NEDC, where mean velocity amounts only 17.8 km/h, the required bottle volumes can be less as shown above. The same bottle volumes are needed for full combustion and pneumatic engine operation, because at 3500 rpm, the engine consumes 0.12 g/cycle. Obtaining of high performance for four-cylinder engine fed by the pressurized air is possible with the total air consumption 0.48 g/cycle. At this rotational speed for only 5 min of the drive, only 4.2 kg of air is needed, which requires the bottle volume about 10 l.

9. Summary

Across the world, some universities and companies successfully manufactured different types of air powered engines (APEs). But generally limited by low working efficiency of the compressed air, low-temperature ice block and critical parts' performance, the APE is still in the development stage. Therefore, precise modelling and simulation cannot only lay a foundation for the design of the APE but also save development costs.

1. Two-stroke pneumatic engine enables higher values of bmep at lower air pressure (below 150 bar), but at higher air pressure, the four-stroke engine gives the same bmep but indicates lower value of SAC. A higher compression ratio in the four-stroke engine than in the two-stroke engine forces of applying of higher pressure of the injected air.
2. The two-stroke pneumatic engine indicates smoother work because of existence of expansion process every rotation of the crankshaft and show high power at smaller air pressure, but show also a higher value of SAC. Working parameters of such engine depend mainly on pressure of the injected air, start point and time of duration of air injection, design of the air valve, way of opening of the air valve and other control parameters.

3. Higher injection pressure of the air increases bmep as well as increases the value of SAC. The engine at full load shows a decrease of bmep and an increase of SAC with growing of rotational speed at the same injection pressure.

4. The mean temperature of working medium should be at above 300 K for proper lubrication of the two-stroke pneumatic engine. Because of such requirement, the injection pressure of air should be above 50 bar (see **Figure 4**).

5. The presented hybrid vehicle with the combustion and pneumatic engine enables to achieve higher mean effective power and torque in comparison to the standard engine. The described hybrid engine can work in four modes (Section 6.1) depending on required power and reduction of exhaust gas emission.

6. In the pneumatic working mode, the CI engine does not emit any toxic chemical species (zero emission) mostly in urban drive, which causes also a lower heat exchange with a cooling system (higher thermal efficiency).

7. During the pneumatic working mode, the CI engine demands the air volume equal above 6 l under pressure 350 bar for the driving with constant velocity of 50 km/h in the city with a distance 10 km. For higher driving distance, the foreseen tank volume amounts not more than 30 l. Applying an additional high pressure air injection above 350 bar, one increases the theoretically total efficiency of the engine from 4 to 19% at micro-dose of air depending on the duration of air injection. The engine operating in the air microdose mode of the air can indicate the total efficiency higher than 50%. The engine working at full dose of air (combustion and pneumatic mode) indicates much higher total efficiency reaching value 60%.

8. Additional air in CI engines decreases CO, HC and NOx emission, which is caused by additional oxygen and lowering of temperature in the cylinder. In the compression ignition engine, a lower soot emission is expected, which is caused by penetration of fuel jet core by the injected air. The hybrid vehicle with combustion and pneumatic engine enables to reach zero emission of exhaust gases with high engine performance, despite it requires an additional power for compressor drive, which is recovered during vehicle deceleration.

In this chapter, we presented only chosen aspects of clean pneumatic and hybrid combustion and pneumatic piston engines for application in transportation based on the own works. The presented experimental and theoretical works of other researchers indicate the needs of alternative and more ecological driving sources.

10. Future work

Future works should be concentrated on optimization of the air injection parameters in order to obtain a maximal efficiency. This can be performed also by simulations, but the control parameters should be verified on the experimental stand with real engine. Without the

http://dx.doi.org/10.5772/intechopen.69689

national subsides or financial support from industry, the research work in scientific institutions will not be possible. Now the passenger cars are equipped only with four-stroke engine. For that reason, the research work should be carried out on pneumatic four-stroke engine and hybrid pneumatic and combustion engines of the same type. A wider use of considered engines requires also an engagement of the industry. Research work under development of pneumatic and hybrid engines will be focused on further simulation and adaptation of production engines for the pneumatic feeding and testing them on laboratory stands also in dynamic modes. Correlation of simulation and experimental results will give a more precise model of the pneumatic engine.

Abbreviations

HCCI	Homogeneous charge compression ignition
CAI	Controlled auto-ignition
ATAC	Active thermo-atmosphere combustion
CI	Compression ignition
SI	Spark ignition
CA	Crank angle
TDC	Top dead centre
BDC	Bottom dead centre
ATDC	After top dead centre
BTDC	Before top dead centre
WOT	Wide opening throttle
ECU	Electronic control unit
bmep	Brake mean effective pressure
imep	Indicated mean effective pressure
bsfc	Brake specific fuel consumption
SAC	Specific air consumption
AMPC	Air mass per cycle
Λ	Air excess ratio

Author details

Wladyslaw Mitianiec

Address all correspondence to: wmitanie@usk.pk.edu.pl

Cracow University of Technology, Cracow, Poland

References

[1] Lavy J, Angelberger C. Towards a better understanding of controlled auto-ignition (CAI) combustion process from 2-stroke engine results analyses; 2001; SAE Paper 2001 01 1859/4276

[2] Onishi S, Hong Jo S, Do Jo S, Kato S. Active thermo-atmosphere combustion (A.T.A.C.)—A new combustion process for internal combustion engines; 1979; SAE Paper 790501

[3] http://www.douglas-self.com/MUSEUM/LOCOLOCO/airloco/airloco.htm#pars

[4] Cai ML, Kawashima K, Kagawa T. Power assessment of flowing compressed air. Journal of Fluids Engineering. 2006;**128**:402-405

[5] Chen Y, Liu H, Tao G. Simulation on the port timing of an air-powered engine. International Journal of Vehicle Design. 2005;**38**:259-273

[6] Chen P, Yu X, Liu L. Simulation and experimental study of electro-pneumatic valve used in air-powered engine. Journal of Zhejiang University Science A. 2009;**10**:377-383. DOI: 10.1631/jzus.A0820373

[7] Creutzig F, Papson A, Schipper L, Kammen DM. Economic and environmental evaluation of compressed-air cars. Environmental Research Letters. 2009;**4**:044011 (9pp). DOI: 10.1088/1748-9326/4/4/044011

[8] Papson A, Creutzig F, Schipper L. Compressed air vehicles: Drive-cycle analysis of vehicle performance, environmental impacts, and economic costs. Transportation Research Record Journal of the Transportation Research Board. 2010;**2191**:67-74. DOI: 10.1088/1748-9326/4/4/044011

[9] Huang KD, Tzeng SC. Development of a hybrid pneumatic-power vehicle. Applied Energy. 2005;**80**:47-59. Available from:www.elsevier.com/locate/apenergy

[10] Chen H, Ding Y, Li Y, Zhang X, Tan C. Air fuelled zero emission road transportation: A comparative study. Applied Energy. 2011;**88**:337-342

[11] Huang KD, Tzeng SC, Ma WP, Chang WC. Hybrid pneumatic-power system which recycles exhaust gas of an internal-combustion engine. Applied Energy. 2005;**82**:117-132

[12] Huang Ch, Hu Ch, Chih-Jie Yu Ch, Sung Ch. Experimental investigation on the performance of a compressed-air driven piston engine. Energies. 2013;**6**:1731-1745; DOI: 10.3390/en6031731

[13] Yu Q, Cai M. Experimental analysis of a compressed air engines. Journal of Flow Control Measurement & Visualization. 2015;**3**:144-153. http://dx.doi.org/10.4236/jfcmv.2015.34014

[14] Shen YT, Hwang YR. Design and implementation of an air-powered motorcycles. Applied Energy. 2009;**86**:1105-1110

[15] Wang YW, You JJ, Sung CK, Huang CY. The Applications of piston type compressed air engines on motor vehicles. Procedia Engineering. 2014;**79**:61-65

[16] MDI (Moteur Development International) Available from: http://www.mdi.lu/english/2014%20english.php

[17] http://www.themotorreport.com.au/5732/tata-air-car-powered-entirely-by-compressed-air-blow-me-down

[18] http://www.themotorreport.com.au/5732/tata-air-car-powered-entirely-by-compressed-air-blow-me-down

[19] Franco A, Stan C, Eichert H. Numerical analysis of the performances of a small two-stroke engine with direct injection. In: International Congress & Exposition; Detroit; 1996; SAE Paper 960362, SAE

[20] Look DC, Sauer HJ. Engineering Thermodynamics. Boston: PWS Engineering; 1986

[21] Blair GP. Design and simulation of two-stroke engines: R-161. Warrendale: SAE; 1996

[22] Mitianiec W, Jaroszewski A. Mathematical Models of Physical Processes in Combustion Engines of Small Power. Wroclaw-Warsaw-Cracow: Ossolineum; 1993

[23] Chen CH, Veshagh A. A comparison between alternative methods for gas flow and performance prediction of internal combustion engines; 1996; SAE Paper 921734

[24] Annand WJ. Heat transfer in the cylinders of reciprocating internal combustion engines. Proceedings of the Institution of Mechanical Engineers. 1963;**177**(1):973-996 https://doi.org/10.1243/PIME_PROC_1963_177_069_02

[25] Heywood JB. Internal Combustion Engine Fundamentals. McGraw Hill, New York; 1988

[26] Mitianiec W, Wiatrak W. Pneumatic two-stroke engine as an alternative power source. Journal of Kones Powertrain and Transport. 2008;**15**(3):357-366. European Science Society of Powertrain and Transport Publication; Warsaw

[27] Mitianiec W, Wiatrak W. Study of combustion and pneumatic spark ignition engine. Journal of Kones Powertrain and Transport. 2010;**17**(1):283-290. European Science Society of Powertrain and Transport Publication; Warsaw

[28] Kumar N, Banka U, Chitransh M, Takkar J, Kumar V, Gupta U, Singh S. Compressed air retrofit kit for existing motor vehicles. Proceedings of the World Congress on Engineering. 2013; Vol. **III**, WCE 2013, ISSN: 2078-0966 (Online); London

[29] Xu Q, Shi Y, Yu Q, Cai M. Virtual prototype modelling and performance analysis of the air powered engine. Journal of Mechanical Engineering Science. 2014;**228**(14):2642-2651

[30] Wiatrak W. Hybrid pneumatic-combustion engine. Patent Application No. 384724. Polish Patent Office, BUP; 28.09.2009

[31] Wiatrak W, Mitianiec W. Engine with two different energy sources consumption in one working cycle. Patent Application BUP 22/2012. Polish Patent Office; 22.10.2012

[32] Kadijk G, Ligterink N. Road load determination of passenger car; 2012. TNO 2012 R10237

Chapter 8

Power Cylinder System for Internal Combustion Engines

Chao Cheng

Additional information is available at the end of the chapter

http://dx.doi.org/10.5772/intechopen.69762

Abstract

Piston ring pack is one of the most critical components for engine performance, durability, and emission. It has become a decisive factor for engine life. From previous study, a three-dimensional ring model has been developed using finite element method to study the interactions between the ring-cylinder liner and the ring-groove side interfaces. The ring-cylinder and ring-groove side contacts are modeled using finite element method based on penalty method optimization algorithm. Ring deformation, reaction forces at the ring sides and ring face, and the twist angles along the entire ring circumference are obtained from the model. However, the dynamic behavior of the ring is still less understood. In this study, the dynamic response of the ring over an engine cycle is studied for a second compression ring with a non-symmetric cross section.

Keywords: power cylinder system, piston, ring pack, cylinder liner, friction, wear, oil consumption, dynamics, finite element, optimization, internal combustion engines

1. Introduction

The internal combustion engine converts thermal energy of the combustible fuel into mechanical energy that moves the piston and eventually the crankshaft. This energy conversion process occurs within the engine power cylinder system. The power cylinder system comprises the following components: piston, piston rings, cylinder liner, wrist pin, and connecting rod.

The piston is the main component that delivers the mechanical energy through reciprocating motion. And this reciprocating motion is transmitted into rotational motion of the crankshaft to output power through the connecting rod. The connecting rod's small end is connected to the piston through the wrist pin, and the rod's big end is connected to the crankshaft. Combustion occurs above the piston in the combustion chamber, which is sealed by the ring pack, especially the top compression of the ring pack. **Figure 1** shows these major components of the power cylinder system.

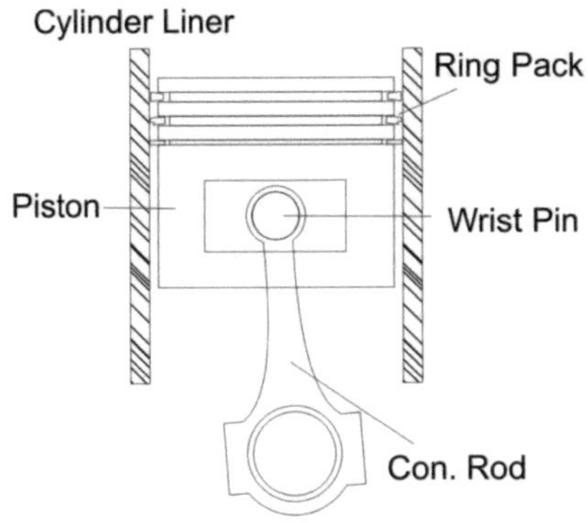

Figure 1. Power cylinder system.

A complete engine cycle consists of four different strokes for a four-stroke engine along with the piston-reciprocating motion. These four strokes are intake stroke, compression stroke, expansion stroke, and exhaust stroke as shown in **Figure 2**.

As for a modern diesel engine, which is known for its better efficiency over its gasoline counterpart, only about 40% of the energy produced by the engine is converted to the engine output

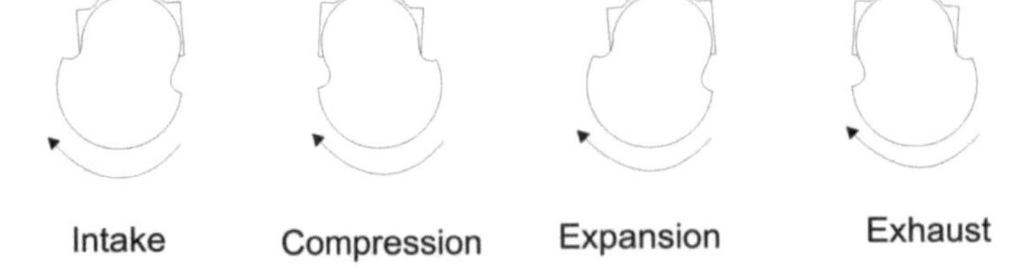

Figure 2. Four strokes for a complete engine cycle.

power. About 4–15% of that energy is wasted as mechanical friction loss. And the rest of the energy, which is almost over half of the chemical energy, is dissipated as other forms, for example, heat transfer, blowby loss, and so on, as shown in **Figure 3** from the study by Richardson [1].

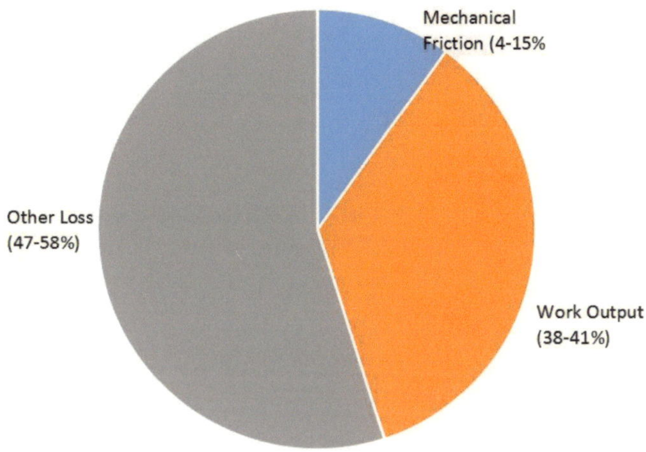

Figure 3. Power distribution for diesel engines.

And about half of the mechanical friction loss is attributed to the friction in the power cylinder system, including the piston, ring pack, and the connecting rod as shown in **Figure 4** [1]. The other part is due to the friction of other components, for example, the valve train system, the crankshaft bearings, and so on.

The friction loss distribution among piston, piston ring pack, and the connecting rod for the power cylinder system can be found in **Figure 5** [1]. As can be found, the piston and ring pack account for higher friction loss than the connecting rod.

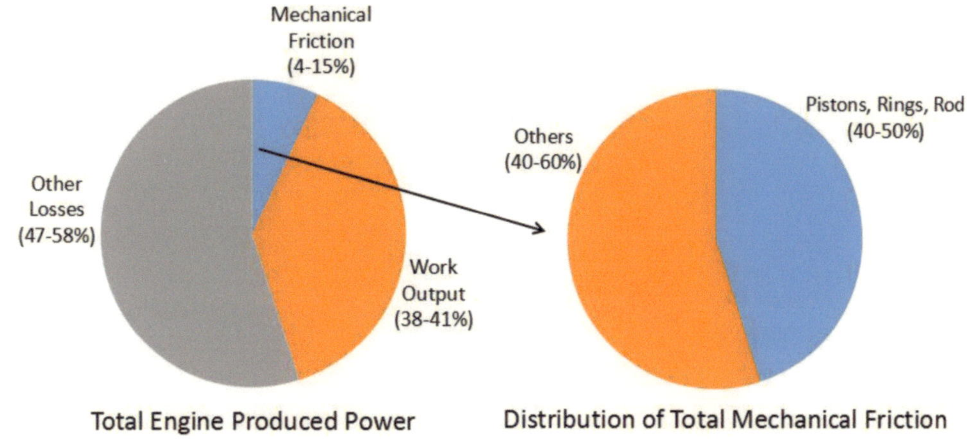

Figure 4. Mechanical friction power distribution.

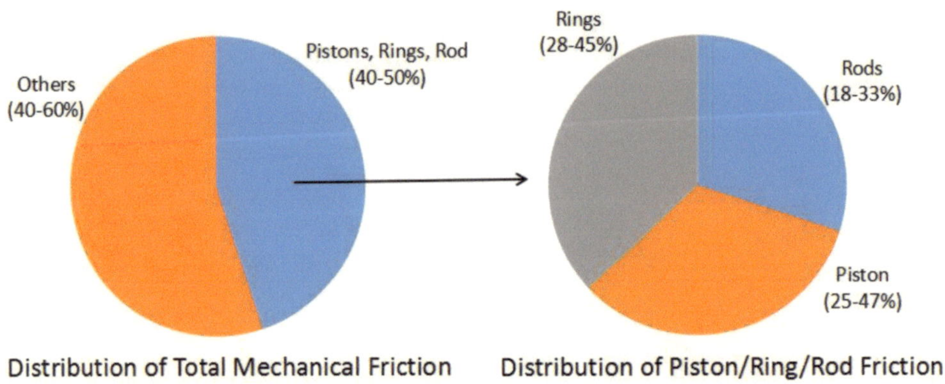

Figure 5. Power cylinder system friction power distribution.

1.1. Piston

The piston of an internal combustion engine is the main component to transmit the thermal energy into mechanical energy. High-pressure gas from the combustion of the fuel-air mixture pushes the piston downward to deliver mechanical energy. Thus, the working condition for the piston is severe. Pistons in small engines are made of aluminum while for large lower speed applications, the pistons are made of cast iron [2]. As the load continues to increase for engines, especially in the heavy-duty industry, steel pistons become widely used nowadays. **Figure 6** shows a typical piston for diesel engine with the definitions of the key geometries shown in **Table 1**.

The piston skirt generally has a barrel/parabolic profile that promotes hydrodynamic lubrication due to the edge effect (**Figure 7**). This skirt profile needs to be optimized in order to minimize piston friction. Piston skirt also grows outward in the radial direction at a high temperature during engine operation.

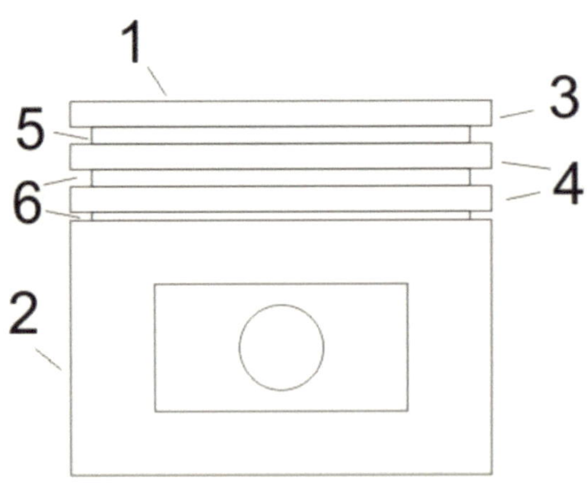

Figure 6. Key piston geometries.

No.	Definitions
1	Piston crown
2	Piston skirt
3	Top land
4	Second and third land
5	Top groove
6	Second and third groove

Table 1. Definitions of key piston geometries.

In addition to the barrel/parabolic profile in the axial direction, piston skirt usually has ovality in the circumference direction as well. The ovality is defined as the difference between the diameter in the thrust axis and the diameter in the pin axis. The ovality is introduced to reduce wear and the risk of scuffing. Development work related to piston dynamics, friction, scuffing, and so on can be found from the references by different researchers [3–11].

1.2. Ring pack

The ring pack is typically composed with three rings: two compression rings and one oil control ring. The main functions of the ring pack are listed as follows:

1. To seal the combustion chamber in conjunction with the piston lands and the cylinder wall, in order to prevent the high-pressure gas from leaking into the crankcase that is wasted in producing power.
2. To control the lubrication oil from getting into the combustion chamber from below the piston as well as to distribute the lubrication oil evenly on the cylinder wall.
3. To transfer heat from the piston to the cylinder wall and eventually to the cooling system. Since the piston crown is exposed to the combustion chamber, it is critical to reduce the piston temperature in order to guarantee the piston's working condition.

Figure 8 shows typical ring packs for modern gasoline and diesel engines.

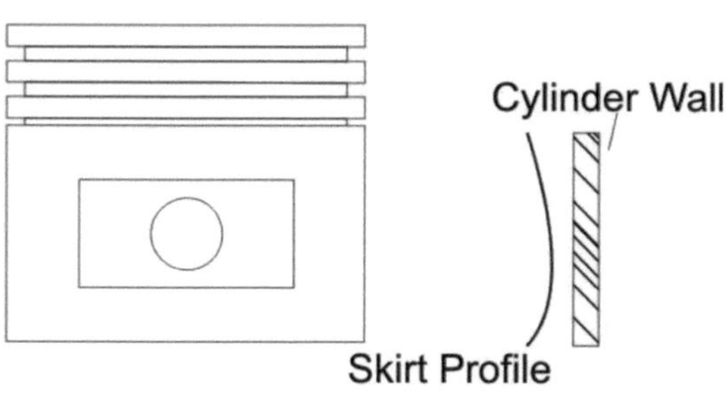

Figure 7. Piston skirt profile.

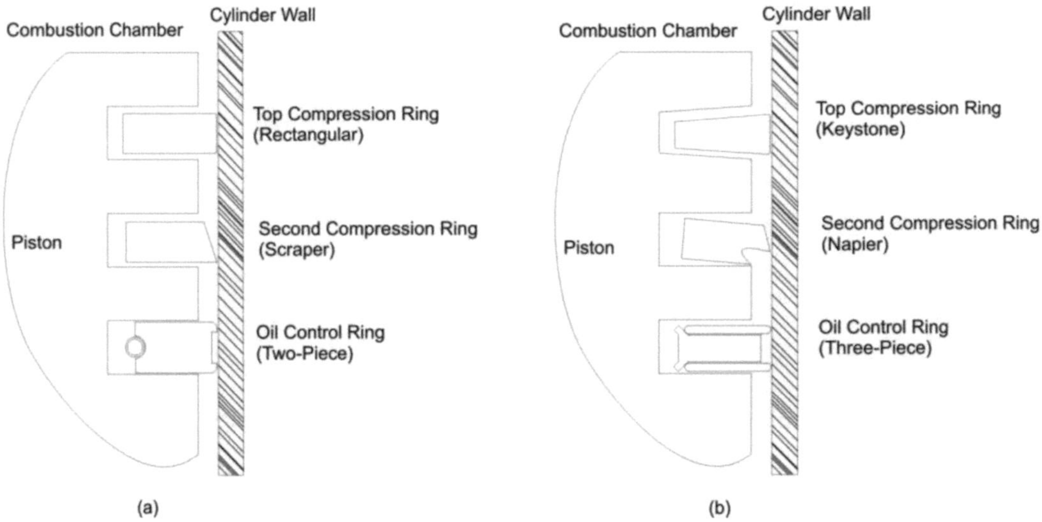

Figure 8. IC engine ring pack: (a) gasoline engine and (b) diesel engine.

1.2.1. Top compression ring

The top compression ring is the first ring and the main component sealing the combustion chamber for engine blowby control. The top ring is also under the most severe working condition since it is directly exposed to the combustion gas and usually under high pressure and high temperature.

The top compression rings for gasoline engine usually have a rectangular cross section. However, for diesel engine operation, the top compression rings usually are keystone rings (**Figure 9**) which promotes the breakup of the deposits between the ring and piston groove, thus reducing the possibility of micro-welding between the piston ring and the piston groove. The top compression ring usually has a parabolic or a barrel profile at its front face in order to enhance the hydrodynamic lubrication between the ring face and the cylinder wall interface (**Figure 9**).

The sealing capability of the top compression ring has significant influence of engine blowby because of the high gas pressure gradient across the top ring. Engine blowby is recognized as the high-pressure gas leaking into the crankcase through the ring pack. Thus, the top compression ring is desired to conform to the cylinder wall evenly along the ring circumference. Also, due to the high gas pressure gradient across the top ring, the top ring stays against the bottom side of the piston groove most of the time during the engine cycle.

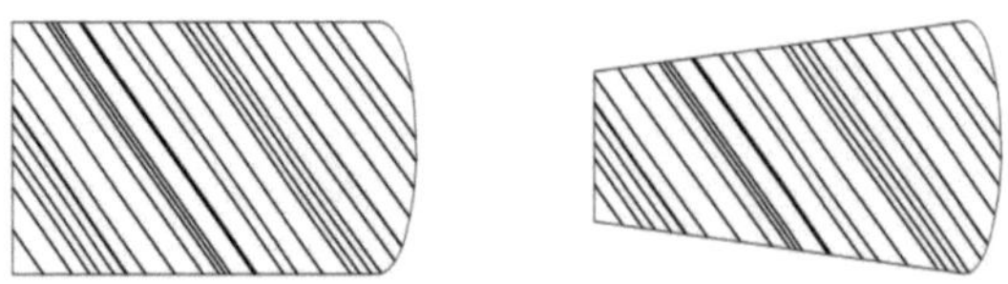

Figure 9. Top compression ring cross section.

1.2.2. Second compression ring

The second ring is a scraper ring which is recognized as 80% for scraping the lubrication oil down and 20% for sealing the combustion chamber. Because of the wedge effect, the scraper ring promotes hydrodynamic lubrication during the up-strokes (compression and exhaust strokes) and scrapes oil down during the down-strokes (intake and expansion strokes). **Figure 10** shows two types of second rings: one is scraper ring and the other is Napier ring. For the second ring, static twist is usually introduced by cutting off the ring material at one of the back corners. If the lower inside corner is cut off, the ring is a negative static twisted ring, while if the upper inside corner is cut off, the ring has a positive static twist configuration.

Although the gas pressure gradient across the second compression ring is much lower than that of the top ring, the second ring also has a noticeable effect on gas flow and gas dynamics. Due to this lower-pressure gradient across the second ring, the ring inertial force becomes competitive to gas pressure force. The inertial force may lift the second ring up at late compression stroke such that the second ring stays against the top flank of the groove. This process may repeat depending on the pressure buildup above the second ring when it is top seated. This unstable axial in-groove motion is recognized as ring fluttering [12]. When the ring fluttering occurs, another gas flow path between the ring and groove sides opens. As a result, blowby gas may increase.

It is also possible for the second ring to move inward in the radial direction. This radial movement is known as ring radial collapse [12]. When the ring radial collapse occurs, the gas above the ring can flow past the ring directly between the ring face and the cylinder wall to the lower land. Severe engine blowby can occur at this ring collapse condition. It depends on the ring and piston design which of the two conditions occur, ring fluttering or ring collapse. It is also possible that the two conditions occur simultaneously.

It was found that the static twist has significant influence on the second ring fluttering and radial collapse. The second ring with a negative static twist is more likely to flutter than a positive static twist second ring. However, if the second ring is lifted against the top flank of the groove, the positive static twist configuration will be more likely to collapse than the negative twist configuration. This will be discussed in the "Ring dynamics" section later in this chapter.

1.2.3. Oil control ring

The oil control ring is used to meter and distribute lubrication oil onto the cylinder wall. There are generally two types of oil control rings: two-piece oil control ring and three-piece oil

Figure 10. Second compression ring cross section.

control ring (**Figure 11**). The two-piece oil control ring consists of a ring body with two rails and a helical spring on the back providing the ring tension force. The three-piece oil control ring consists of two segments and an expander in between the two segments. The expander provides the radial force to conform the ring to the cylinder wall and also the axial force to push the ring against the top and bottom sides of the groove. The oil control ring is a two-direction scraper ring that scrapes oil in both upward strokes and downward strokes. During the downward strokes, the bottom rail/segment scrapes oil directly back into the crankcase. The top rail/segment scrapes oil back into the groove through the oil control ring expander. Generally, holes at the back of the oil control ring groove can be found along the circumference in order to allow the oil draining to the crankcase. In some piston design, instead of using these holes at the back of the groove, cast slots are introduced at the bottom edge of the groove for oil drain as an easier solution. During the upward strokes, the bottom rail/segment scrapes oil into the groove through the expander. The recovery of oil scraped by the top rail/segment during these upward strokes depends on the external force on the top rail/segment. At times, the external axial force on the oil control ring overcomes the expander force. As a result, an oil flow crevice is formed between the oil control ring and the groove sides allowing the oil drain into the groove and eventually back to the crankcase.

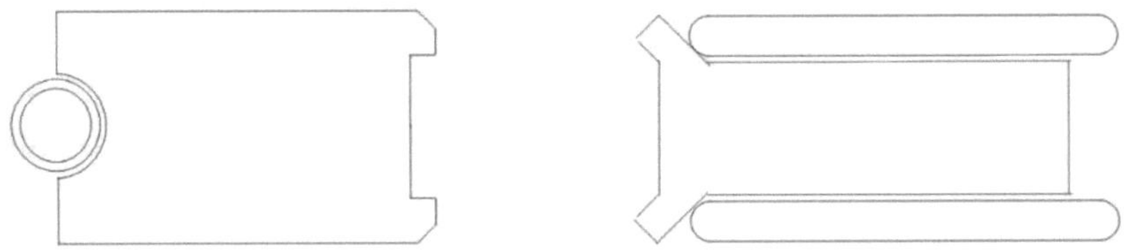

Figure 11. Oil control ring: two-piece oil control ring (left), three-piece oil control ring (right).

1.3. Cylinder

The cylinder of a reciprocating engine is the part through which the piston travels. The cylinder may be sleeved or sleeveless depending on the metal used for the engine block. For example, a cast iron engine block generally does not require cylinder sleeve because the iron is hard enough to resist wear between the piston ring and the cylinder wall. However, for aluminum alloy engine blocks that can be found in almost all daily drive cars, cylinder sleeves are required since the aluminum alloy is not hard enough to resist wear between the piston ring and the cylinder wall interface.

Cylinder liners, or cylinder sleeves, are manufactured nowadays using the centrifugal casting process. The centrifugal casting process refers to the technique for casting, which has a permanent mold spinning continuously along its center line at a constant speed. At the same time, molten metal is poured to the mold and thrown toward the inside wall of the mold. Then, the molten metal is solidified after cooling. The spinning orientation of the casting machine can be either horizontal or vertical, depending on the parts it is producing. Horizontal spin is preferred for long and thin cylinder, while vertical spin is preferred for short and wide cylinders. Aluminum engines without sleeves can also be found. The aluminum cylinders are treated with nickel silicone alloy coating or other plasma coating that help reduce cylinder wear. Other techniques have also been explored by the researchers in order to reduce engine friction. One method is to introduce dimples at the mid-stroke to the cylinder walls [13]. This helps reduce friction because at the mid-stroke, the piston rings are generally under hydrodynamic friction when the piston speed is high. By introducing the dimples to the cylinder wall, the effective area of contact between the ring faces and the cylinder wall has been reduced. This leads to reduction of viscous friction as claimed.

Typical surface roughness for cylinder liner is 0.4–0.5. This roughness has been reduced significantly, which could help reduce engine oil consumption. Rougher cylinder walls can help retain lubrication oil on the liner surface between micro-valleys, which is similar to the dimple liner [13]. As a result, friction between the ring/cylinder wall and the piston skirt/cylinder wall interfaces can be reduced due to the lubrication oil in the micro-valleys. However, this micro-valley-retained oil is not scraped from the liner during engine down-strokes and can stay exposed to high-temperature gases. As a result, more oil is evaporated and the oil consumption increases.

Cylinder liners are no longer circular when the engine is in operation. The deformation results from mechanical distortion from bolting the cylinder block to the cylinder head, thermal distortion when the thermal load on the liner is not uniform, mechanical load when piston is slapping into the liner, the pressure load from the combustion event, and so on. Cylinder bore distortion is measured from an experiment by researchers [14]. For modeling concern, the cylinder bore distortion is usually defined by a Fourier series [4, 5]:

$$\delta R = \sum_{i=0}^{i=4} \left(A_i \cos(i\theta) + B_i \sin(i\theta) \right) \tag{1}$$

where δR is the deviation from roundness, A_i and B_i are Fourier coefficients and i is the order of the series.

The orders of the distortion are recognized in **Table 2**.

Zero order	Change in bore diameter
First order	Bore eccentricity
Second order	Oval deformation
Third order	Three-lobe deformation
Fourth order	Four-lobe deformation

Table 2. Cylinder bore distortion.

2. Ring pack dynamics

The piston ring dynamics is closely related to their functions, especially for gas control and oil control. Although the top ring is the most important part for gas sealing, while the oil control ring has the highest effect in control oil flow and consumption, the second ring also has significant influence on both gas and oil control. This section discusses the ring dynamics of the second compression ring. The theories can also be applied on the top compression ring and oil control ring as well and the details of the ring dynamics models can be found from Refs. [15–20].

As discussed in Section 1, there are two types of ring dynamics: ring fluttering and ring radial collapse. Piston ring fluttering is the axial movement under the consequence of external force unbalance, especially between gas pressure force and inertial force. The other loads acting on the ring, including friction force, oil film squeezing force, and so on, are relatively small in comparison [6]. It is noted that while the second ring friction is relatively low, the friction forces of the oil ring and the top ring during high cylinder pressures can be large. Also, only the second ring flutter and collapse that occurs around top dead center (TDC) firing conditions is described here. This region is also considered the most important region for ring fluttering and collapse because of its significance on blowby and oil consumption.

Another phenomenon, radial collapse, can occur if the ring is lifted and seated against the top of the ring groove. When the ring is on the top side of the ring groove, the pressure force not only pushes the ring downward but also acts on the front face of the ring pushing it inward. The ring seals the gas pressure at the top, meaning the pressure behind the ring can be much lower. When the pressure force on the ring face exceeds the ring tension and the pressure force behind the ring, the ring collapse will occur. Once it collapses, the gases will escape past the ring face and equalize all around the rings. Once again, there will be no net gas pressure on the ring and the elastic tension of the ring will force the ring back out to the cylinder wall. As can be expected, there is no sealing between the ring face and the cylinder wall. As a result, the gas flow can pass the ring face resulting in high blowby. The ring collapse is one of the unstable behaviors of the ring.

It will depend on the ring and piston design as well as the operating conditions if ring flutter or collapse can occur. It is also possible that both ring flutter and collapse occur at the same time. In either case, the second ring loses its sealing capability allowing gases flowing either around the ring (in the fluttering case) or past the ring face (in the ring collapse case).

The ring design itself also has significant influence on its stability, for example, ring static twist. A negative-twisted second ring forms an outer edge seal between the ring and groove bottom sides when the ring is at the flank bottom. This allows gases to flow underneath the ring resulting in a very low net downward gas pressure force. In this case, the ring can be easily lifted by the inertial force acting on the ring (**Figure 12a**). On the other hand, for a positive static twisted second ring, the seal between the ring and the groove bottom occurs at the inner bottom corner. This prevents higher-pressure gas moving between the ring bottom and the groove bottom, resulting in a higher downward pressure force. The ring is not easy to be lifted by inertial force. **Figure 12a** depicts a simplistic illustration showing the gas pressure forces acting on the sides of the rings.

Similarly, the ring top-seating stability (**Figure 12b**) can be explained similarly as for the bottom-seating condition. However, it should be noted that since a negative-twisted ring is easier

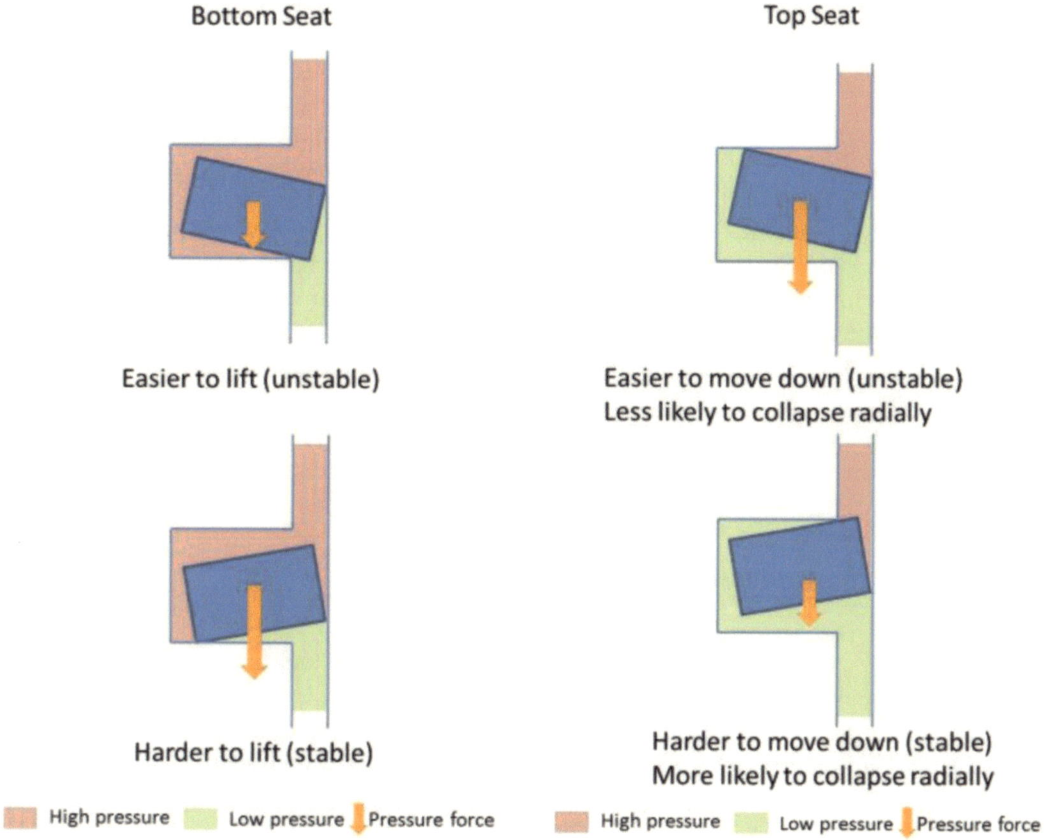

Figure 12. Ring-seating stability: (a) bottom-seating stability and (b) top-seating stability.

to move down, it will be less likely to collapse. Conversely, the positive-twisted ring will be more difficult to push down; therefore, the ring will be more likely to collapse radially inward as the pressure above the ring can become higher. In summary, the configuration with positive static twist tends to increase the pressure force holding the second ring down and promoting second ring stability. This positive twist ring is also more susceptible to collapse. On the contrary, the configuration with a negative static twist will tend to promote ring flutter. However, ring radial collapse is less susceptible to occur.

When the second ring flutters or collapses, the blowby will generally be higher. This is because the ring does not seal the gases and the gases flow past the ring. While this can cause high blowby, the pressure in the second land will be very low. This will prevent reverse blowby, which is beneficial for oil consumption. More discussions about ring pack dynamics can be found from Refs [17–20].

Nowadays, researchers from the industry and the academic area are developing the ring pack dynamics model in three-dimensional (3D) in order to capture the variation along the ring circumference with the consideration of cylinder liner ID deformation. In addition, the influence from piston secondary motion can also be implemented to further understand the ring pack behavior. This will capture the gas flow in the circumference direction, which the current commercial two-dimensional (2D) models are not capable of. As a result, the ring dynamics, oil consumption, friction, and wear for the ring pack can be better modeled and understood to guide design. The next section is an introduction to the 3D modeling work for the ring pack.

The 2D ring pack dynamics model is still widely used in the automotive and heavy-duty industries during product development, given the experience and fidelity built on this approach. Some improving activities include implementing wear model at the ring face and side based on different wear mechanism, oil consumption model due to oil evaporation, oil throw-off, oil scraped back to the combustion, and so on. In addition, 3D ring pack dynamics models are being developed using different approaches, including full FEA with hexahedron element, discretizing the ring using space beam elements, and so on, with different orders of success. The 3D model approach will be discussed in the next section with more detail.

3. Ring-piston groove-cylinder liner interaction

In engine power cylinder system development, utilizing CAE tool has become a standard approach to design and optimize the system. Traditional CAE tools are two-dimensional (2D) which considers the ring motion along the cylinder axis and the twist. However, the variation along the ring circumference is assumed to be identical. The demand for better understanding of the power cylinder system requires three-dimensional (3D) CAE tools to simulate the variation along the ring circumference as well. Researchers have started working on 3D modeling. One variation along the ring circumference is the contact pressure between the ring face and the cylinder bore interface, and the ring side and piston groove side interface. The interactions are discussed in this section.

3.1. Ring-cylinder bore contact

When the free-state ring is installed into the cylinder liner, the ring is constrained at its front face by the cylinder wall. Every point on the ring front face needs to be tracked whether it is in contact with the cylinder wall or not. However, due to the computation time and resource, it is not possible with the existing computation tool. And the most important thing is how the contact force/pressure distributes along the ring circumference. Thus, in this section, the ring is specified constrained at 13 different cross-section locations along the circumference [21–23]. The ring conformability is modeled using finite element method (FEM) [24, 25] for a keystone compression ring. The approach solving the problem is based on penalty method-based optimization that minimizes the strain energy of the piston ring [26–30].

As shown in **Figure 13**, the middle constraint locates at the ring back (opposite to the ring gap) front face. Other constraints are symmetric about the ring back and distribute with an increment of about 30°. The free-shape ring mesh and the deformed ring mesh without temperature compensation are shown in **Figure 13**.

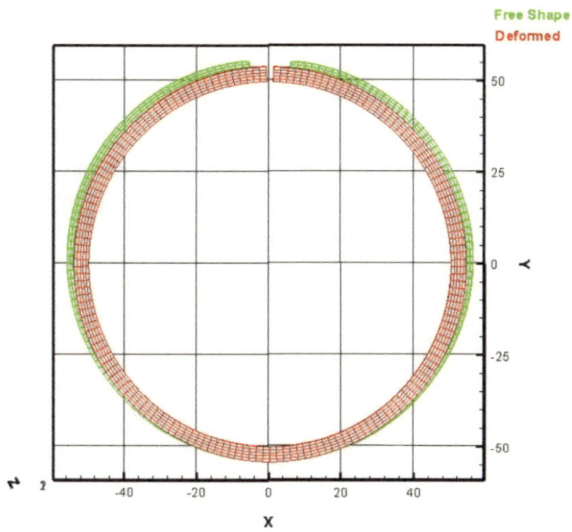

Figure 13. Free-shape and deformed ring meshes.

The green mesh shown in **Figure 13** represents the free-shape ring while the red mesh represents the deformed ring shape under the cylinder bore constraints without temperature compensation. It is obvious that the ring is pushed inward from its free state. The constraint forces that push the ring to its deformed position are shown in **Figure 14**. The blue and red bars represent the constraint forces at a certain circumference location at the upper and lower corners at ring face. And the green and purple dots show the separation gaps between the ring face and the cylinder bore.

From **Figure 14**, it is found that the two contact forces at the same cross section are identical since the ring has a symmetric cross section and there is no twisting moment on the ring. The

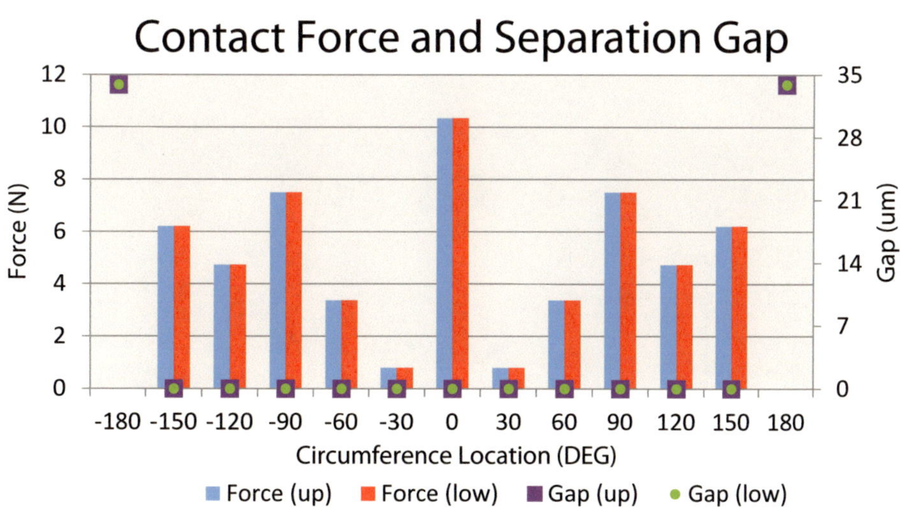

Figure 14. Constraint force and separation gap.

plot also shows that the constraint force at the ring back is the highest. At the cross sections approximately 30° away from the ring back, the lowest constraint forces are found for the sections that are in contact against the cylinder wall. The constraint forces at the ring tips vanish such that the ring separates from the cylinder wall in its front face at its two tips. The separation gap is defined as the radial distance between the cylinder wall ID and the ring tip OD. A 34-μm separation gap is found for this specific ring from the FEA model.

3.2. Result of ring-cylinder bore-groove side contact

Another example is given in this section for ring-cylinder bore-groove side contact using a scraper ring with a positive static twist. The scraper ring has a taper face and cuts off at the ring inner upper corner, which promotes positive twist when installing the ring into the piston groove. The cross section of the scraper ring is shown in **Figure 15**.

From **Figure 15**, four nodes of the cross section at a given circumference location are considered for the ring-piston groove side interaction and are numbered as node 1, node 2, node 3, and node 4 as shown. These four nodes are constrained by the groove in the axial direction. This means nodes 1 and 2 should stay in contact or above the groove bottom side, while nodes 3 and 4 should stay in contact or below the groove top side. Two nodes on the ring front face are constrained by the cylinder bore in the radial direction, at the front face top and bottom edges, respectively. The groove has zero angles at its top and bottom sides. The nominal clearance between the groove and the ring axial thicknesses is 0.1 mm.

The main parameters describing the ring are listed in **Table 3**.

The constraint locations along the ring circumference are equally spaced with about 30° from one butt end to the other. The number of constraint locations is found to be able to represent the ring/cylinder liner/groove side contact force/pressure distribution pattern and also save calculation time. Increasing constraint locations will increase computation time exponentially, while decreasing the constraint locations may result in the contact force/pressure pattern not being able to be well represented.

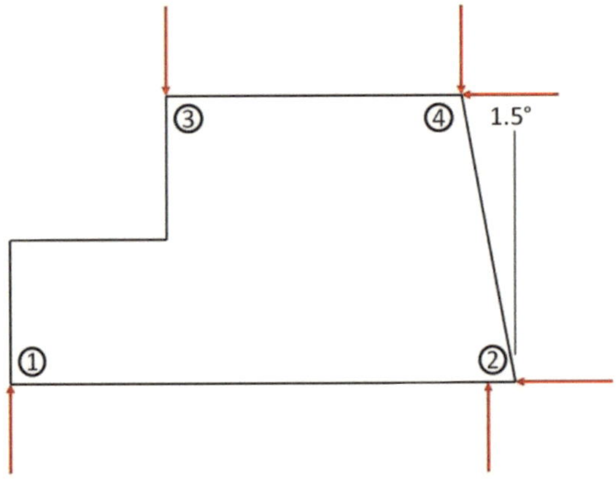

Figure 15. Constraints on ring cross section.

Ring material	Steel
Modulus of elasticity	200.0 GPa
Poisson's ratio	0.3
Cylinder bore diameter	108.0 mm
Coefficient of thermal expansion	13.0E-6/°C
Thermal conductivity	45 W/m K
Ring/gas-convective coefficient	25 W/m² K
Ring/oil film-convective coefficient	100 W/m² K

Table 3. Main parameters for the ring.

The deformed ring shape is shown in **Figure 16** after installing into the cylinder liner and piston groove. The displacement in the *z*-direction (axial direction) is amplified by 100 times in order to illustrate the ring deformation distinctly.

In this case, the ring back and ring butt ends are in contact with the groove bottom side, while the ring touches the groove top side at about 60° from the end gap (120° from the ring back). The constraint forces between the ring and the piston groove sides are important since it dictates the contact pattern which will affect the ring-groove side wear eventually. More details about the ring, cylinder liner, and piston groove interactions can be found from Refs. [22, 23].

Ultimately, the interactions between the ring face-liner bore interface and the ring side-piston groove side interface are used to model the wear between them [17] as well as the ring pack dynamics that heavily influences engine oil consumption, to further optimize the ring pack and power cylinder design and improve the durability of the subsystem.

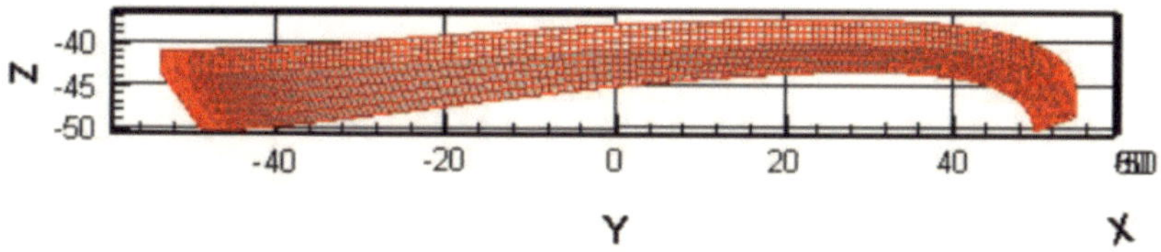

Figure 16. Deformed ring shape after installing into the cylinder liner and piston groove.

Author details

Chao Cheng

Address all correspondence to: chengc22@msu.edu

Michigan State University, MI, USA

References

[1] Richardson DE. Review of power cylinder friction for diesel engines. Journal of Engineering for Gas Turbines and Power. 2000;**122**(4):506-519

[2] Heywood JB. Internal Combustion Engine Fundamentals. New York, NY: McGraw-Hill; 1988

[3] Patel P, Mourelatos Z, Shah P. A comprehensive method for piston secondary dynamics and piston-bore contact. In: SAE World Congress; Detroit, MI. 2007. 2007-01-1249

[4] Keribar R, Dursunkaya Z, Ryan J. A Comprehensive Model of Piston Skirt Lubrication. SAE Technical Paper 920483, International Congress & Exposition, Detroit, Michigan; February 24-28, 1992

[5] Wong V, Tian T, Lang H, Ryan J, Sekiya Y, Kobayashi Y, Aoyama S. A Numerical Model of Piston Secondary Motion and Piston Slap in Partially Flooded Elastohydrodynamic Skirt Lubrication. SAE Technical Paper 940696, International Congress & Exposition, Detroit, Michigan, February 28 - March 3, 1994

[6] Cheng C, Akinola A. Piston friction reduction by reducing piston compression height for large bore engine application. In: SAE World Congress, Detroit, MI. 2017. SAE 2017-01-1044

[7] Kim K, Godward T, Takiguchi M, Aoki S. Part 2: The Effects of Lubricating Oil Film Thickness Distribution on Gasoline Engine Piston Friction. SAE Technical Paper 2007-01-1247

[8] Kim K, Shah P, Takiguchi M, Aoki S. Part 3: A Study of Friction and Lubrication Behavior for Gasoline Piston Skirt Profile Concepts. SAE Technical Paper 2009-01-0193

[9] Westerfiled Z, Totaro P, Kim D, Tian T. An Experimental Study of Piston Skirt Roughness and Profiles on Piston Friction Using the Floating Liner Engine. SAE Technical Paper 2016-01-1043, SAE World Congress and Exhibition, Detroit, Michigan, Apail 12-14, 2016

[10] Kobayashi T. Prediction of Piston Skirt Scuffing via 3D Piston Motion Simulation. SAE Technical Paper 2016-01-1044, SAE World Congress and Exhibition, Detroit, Michigan, April 12-14, 2016

[11] Maurizi M, Hrdina D. New MAHLE Steel Piston and Pin Coating System for Reduced TCO of CV Engines. SAE Int. J. Commer. Veh. 2016;**9**(2):270-275.

[12] Cheng C, Schock H, Richardson D. The Dynamics of Second Ring Flutter and Collapse in Modern Diesel Engines. ASME J. Eng. Gas Turbines Power. 2015;**137**(11):111504-111504-8

[13] Urabe M, Takakura T, Metoki S, Yanagisawa, M, et al. Mechanism of and Fuel Efficiency Improvement by Dimple Texturing on Liner Surface for Reduction of Friction between Piston Rings and Cylinder Bore. SAE Technical Paper 2014-01-1661, SAE World Congress and Exhibition, Detroit, Michigan, Detroit, Michigan, April 8-10, 2014

[14] Bird L, Gartside R. Measurement of Bore Distortion in a Firing Engine. SAE Technical Paper 2002-01-0485, SAE 2002 World Congress, Detroit, Michigan, March 4-7, 2002

[15] Tian T, Rabute R, Wong V, Heywood J. Effects of Piston-Ring Dynamics on Ring/Groove Wear and Oil Consumption in a Diesel Engine. SAE Technical Paper 970835, SAE International Congress & Exposition, Detroit, Michigan, February 24-27, 1997

[16] Akalin O, Newaz GM. Piston ring-cylinder bore friction modeling in mixed lubrication regime: Part I—analytical results. Journal of Tribology. 1999;**123**(1):211-218

[17] Baker C, Rahmani R, Karagiannis I, Theodossiades S, Rahnejat H, Frendt A. Effect of Compression Ring Elastodynamics Behaviour upon Blowby and Power Loss. SAE Technical Paper 2014-01-1669. 2014

[18] Poort M, Cheng C, Richardson D, Schock H. Piston Ring and Groove Side Wear Analysis for Diesel Engines. ASME J. Eng. Gas Turbines Power. 2015;**137**(11):111503-111503-9

[19] Westerfield Z, Liu Y, Kim D, Tian T. A Study of the Friction of Oil Control Rings Using the Floating Liner Engine. SAE Int. J. Engines. 2016;**9**(3):1807-1824

[20] Ejakov M, Schock H, Brombolich L, Carlstrom C, Williams R. Simulation analysis of inter-ring gas pressure and ring dynamics and their effect on blowby. ICE-Vol. 29.2, ASME 1997 Fall Technical Conference

[21] Tomanik E. Improved Criterion for Ring Conformability Under Realistic Bore Deformation. SAE Technical Paper 2009-01-0190, SAE World Congress, Detroit, Michigan, April 20-23, 2009

[22] Cheng C, Kharazmi A, Schock H, Wineland R, Brombolich L. Three-Dimensional Piston Ring-Cylinder Bore Contact Modeling. ASME J. Eng. Gas Turbines Power. 2015;**137**(11): 111505-111505-10

[23] Cheng C, Kharazmi A, Schock H. Modeling of piston ring-cylinder bore-piston groove contact. In: SAE World Congress; Detroit, MI, 2015. SAE 2015-01-1724

[24] Fish J, Belytschko T. A First Course in Finite Elements. Chichester, England; Hoboken, NJ: John Wiley & Sons Ltd.; 2007. pp. xiv, 319 p., 318 p. of plates

[25] Cook RD. Finite Element Modeling for Stress Analysis. John Wiley & Sons, Inc.; 1994

[26] Song X, Diaz AR, Benard A, Nicholas JD. A 2D model for shape optimization of solid oxide fuel cell cathodes. Structural and Multidisciplinary Optimization. 2012;**47**:453-464

[27] Song X, Diaz AR, Benard A. A 3D topology optimization model of the cathode air supply channel in planar solid oxide fuel cell. In: Proceedings of the 10th World Congress on Structural and Multidisciplinary Optimization; May 19-24; Orlando, FL, USA. 2013

[28] Panayi AP, Diaz AR, Schock HJ. On the optimization of piston skirt profile using a pseudo-adaptive response surface method. Structural and Multidisciplinary Optimization. 2009;**38**:317

[29] Yu J, Dong X, Wang W. Prototype and test of a novel rotary magnetorheological damper based on helical flow. Smart Materials and Structures. IOP Publishing Ltd.; 2016;**25**(2).

[30] Dong X, Yu J, Yang M. Optimization and experimental study of magneto-rheological fluid damper considering temperature effects. Journal of Vibration and Shock. 2016;**35**(8)